Global Design Trends in Architecture Strip Stacking Architecture 2

全球建筑设计风潮 2

条形 叠式建筑

ThinkArchit 工作室 主编

華中科技大學出版社
http://www.hustp.com
中国 · 武汉

Preface 前言

"Architecture today, more than ever, can be described as pluralistic, varied, diverse, individual or even narcissistic, often based on cultural, climatic or location influences, but at times based on formal expressionism – the need to be unique, to stand out, and the misused notion that every building needs to be an iconic object, a recognizable symbol. Architecture has become the fulfillment of identity through uniqueness with seemingly each project vying for attention."

This statement is meant to be thought-provoking. Browsing these pages and projects, one has to question the motivation for the formal compositions and geometries expressed. What are the goals for these projects? What causes the responses seen in these images? These are reasonable questions to ask of each. As individual objects, each of these projects can be seen as elegant and notable in their own right. They are the epitome of architecture today – noteworthy, recognizable, and thought-provoking.

The projects here are global in location, but international in execution and identity. Architects today are from one place, but are not necessarilyarchitects of the place they are from. They work globally, they think globally, yet their projects often cannot be identified with the place that they are built. Perhaps in our globally-connected world, this is as it should be. Perhaps it is due to the industry that constructs these visions. Perhaps it is the very nature of our global economy.

Together, the projects presented in this book represent a series of executed concepts in form, materiality, composition, and vision across many scales and building types, from urban to garden, from tall to small. Collectively, they provide a measure for global architecture today. They may be disliked by some or revered by others. They may be whimsical or anticipating the future. Some are obviously ordered; some are seemingly chaotic. Some are static, others dynamic. Some test our understanding and concepts of what is normal. Some explore ideas of materiality. Some think in tested ideas of light and shadow. Many rethink technology, using materials and techniques in new and sometimes uncharted ways. Some use digital parametric capabilities as a way of envisioning and executing. The project teams whose work is represented here deal with the real issues of architecture and urbanism today: the buzz words such as green, sustainable, connectedness, net zero, and new urbanism. They are sometimes trendy, looking for immediate recognition, but underlying some of these is the real potential to become ascendant.

Global Design Trends in Architecture is as much about the exciting, new, imaginative, and innovative projects of the last year as it is meant to elicit a thought process about the state of architecture, its impact on our collective environment, and the question of what role architecture should play in our environment, in our cities, and in our lives.

In an effort to facilitate our need for order and to aid our understanding, this book has been loosely divided into typological categories: STACKING ARCHITECTURE, STRIP ARCHITECTURE, IRREGULAR ARCHITECTURE, and FLUID ARCHITECTURE.

Constant looking, understanding, and self-criticism allow us to continue the dialogue and thought processes used to establish our built environment. The goal of this work is to facilitate these vital activities.

Enjoy.

“比之往昔，今日建筑更为多元、多样、丰富，更加个性甚至自恋，经常受到文化、气候或地理位置的影响，而有时又是基于形式上的表现——渴望与众不同，有时甚至有一种误用的概念，认为每个建筑都应该成为地标性的、可识别的象征符号。建筑极力通过独特质地彰显其身份，每个项目都争着夺人眼球。”

这一陈述发人深思。翻看本书中的各个项目，人们可能要问：这些形式组成和几何形态的表现意图是什么？这些图像将产生怎样的效果？就算单独提出来，这些问题都非常合理。作为个体，每个项目都是优雅的。它们是今日建筑的缩影——引人注目、可识别且引人深思。

这些项目遍布世界各地，而在建设及身份上，他们都是国际性的。现在，建筑师往往并不是来自于项目所在地。他们在全球工作、思考，其项目却经常不能通过其所在地进行识别。在联系紧密的世界里，好像就应该是这样的，也许正是因为建筑行业塑造了这样的一些视野；也许这正是全球经济的本性使然。

本书中的项目代表了一系列理念，这些理念通过形体、材料、结构、视觉效果使各种尺度的建筑成为现实。建筑类型从都市到花园式，从宏大到小巧都有。总的来说，他们为今日全球建筑提供了一个衡量尺度。这些建筑或是不被人接受，或是受到人尊敬。他们可能很俏皮，也可能预示着未来的趋势。有些次序井然，有些则看上去凌乱，有些是静态的，有些则是动态的；有些检验我们对于“什么是正常的”的理解，有些对物质性做出了探索，有些则表达了对光影的思考。通过新的、偶尔非常规的方式运用材料，很多建筑重新表达了对技术的探索。有些使用数字参数来视觉化和实施项目。书中代表性项目团队是在处理今日建筑和都市生活的真实问题，即如绿色、可持续、连通性、零耗能和新都市主义等时髦话题。它们有时很时尚，追寻即刻的认同，所暗含的是其真正的上升潜力。

全球建筑设计潮流体现在过去一年那些振奋人心的、充满想象力的创新项目中，同时也必然引发一场思考，关乎建筑现状、建筑对集体环境的影响，以及建筑在环境、城市和人们的生活中应该扮演怎样的角色等问题。

为了方便读者理解，本书大致将建筑分为叠式建筑、条形建筑、不规则建筑和流体建筑。

不断地审视、理解和自我批评，让我们能继续这样的对话和思考，继而得以塑造我们的建筑环境。这些工作的目的是，简化这些对话和思维过程。

开始享受吧！

GORDON R. BECKMAN, AIA
PRINCIPAL & DESIGN DIRECTOR
高登. 贝克曼
美国建筑师协会会员(AIA)
高级主管，设计主任

Contents 目 录

叠式建筑

Index 索 引

办公建筑

文化建筑

商业建筑

居住建筑

教育建筑

交通建筑

科研建筑

医疗建筑

市政建筑

LOUIS QUATORZE
cta
東横イン

条形建筑

条形建筑指的是建筑平面或立面的外形为长方形或正方形，建筑的外表和空间相对有秩序。建筑结构整齐、统一，建筑外观简洁、具有线性是条形建筑的主要特点。

条形建筑

文化建筑

Claire and Marc Bourgie Pavilion of Quebec and Canadian Art

加拿大魁北克特色艺术收藏品博物馆

Design Company / 设计事务所:
Provencher Roy + Associés Architectes
Location / 地点:
Canada （加拿大）
Area / 面积:
5 483 m^{2}
Photography / 摄影:
Tom Arban, Alexi Hobbs, Marc Cramer, Jean-Guy Lambert

视觉亮点

建筑干净、具有现代风格的外观与周边古老的建筑建立了和谐的"对话"。

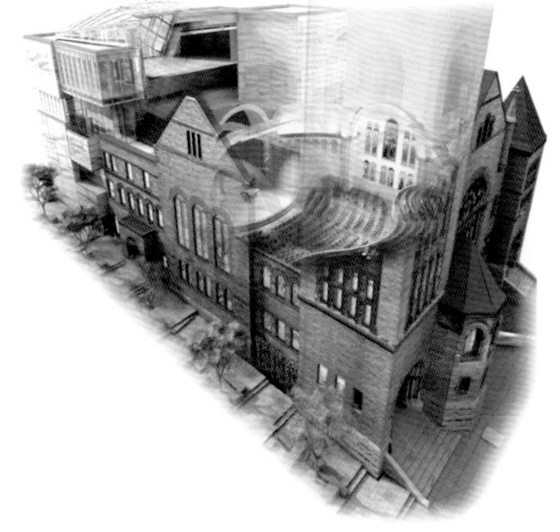

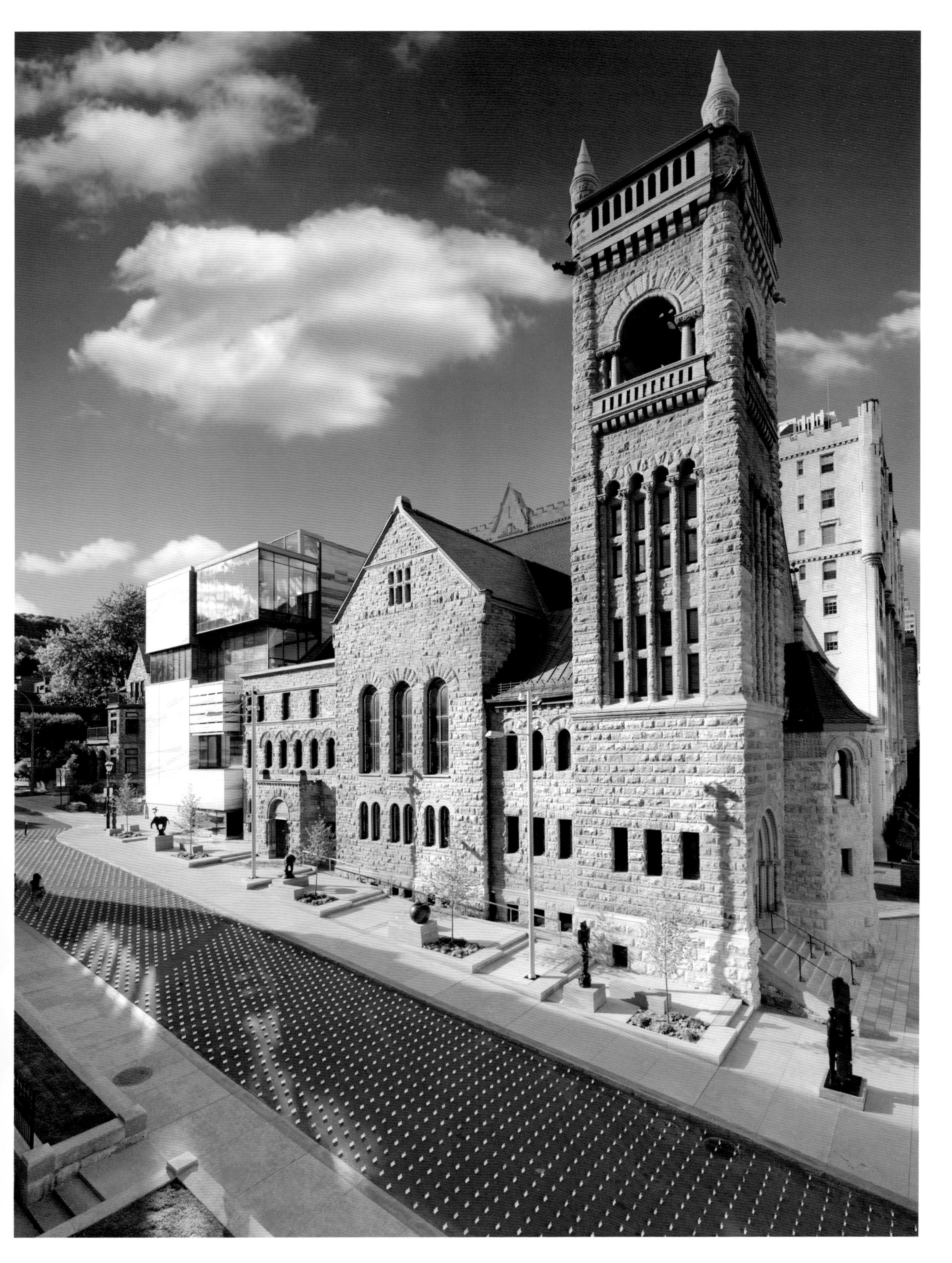

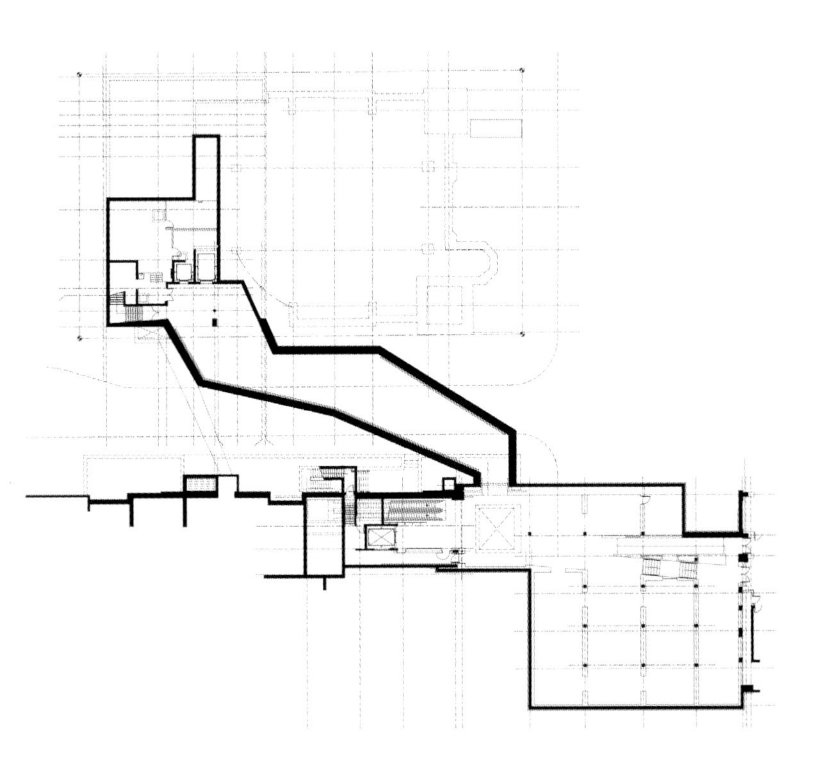

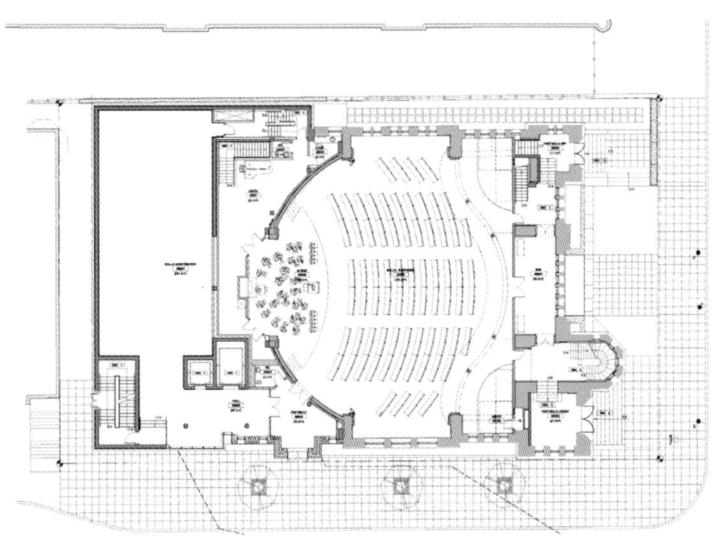

Beside the restored former Erskine and American Church, transformed into a 444-seat concert hall, the rear annex has been completely reconstructed in contemporary style to form the new art pavilion. "The project was complex because it entailed designing a building capable of featuring the Quebec and Canadian art collections while establishing a dialogue with the church, with the museum's other pavilions, and with the city," explained Claude Provencher, founding partner of Provencher Roy + Associés Architectes.

With its restraint and permeability, the new Claire and Marc Bourgie Pavilion of Quebec and Canadian Art establishes a natural dialogue with the city. From every level, the glazed openings offer a view of the city and, at the building's foot, the museum's sculpture garden, a linear exhibition of works of public art bordering the museum. In addition, the glassed-in atrium at the top of the pavilion offers a strong visual link with Mount Royal, an emblematic element of Montreal's identity. Another dialogue is established with the church. The new pavilion shares more than its entrance and reception areas with the former religious building.

这座新建博物馆毗邻一座老教堂，老教堂被改造成一座可容纳 444 人的音乐厅。用于收藏加拿大艺术品的新建博物馆位于老教堂的后部。这座全新的艺术博物馆为现代风格建筑。该项目的一位首席设计师曾这样阐释过该博物馆的设计理念：“该项目的复杂性在于建筑体既要用作展示魁北克与加拿大艺术品，又要在老教堂、博物馆和整座城市之间建立一种联系。”

设计师在新建的博物馆与整座城市之间打造了一种很自然的“对话”。每个楼层均拥有玻璃窗口结构，使人们可以看到城市风景。建筑外部空间设有博物馆雕塑花园。除此之外，博物馆顶部玻璃结构在建筑本身与皇家山之间建立了强烈的视觉联系，皇家山是蒙特利尔市的标志物。另外一种“对话”建立在博物馆与老教堂之间。新建博物馆不仅与老教堂共享入口、接待区，还共享一些其他空间。

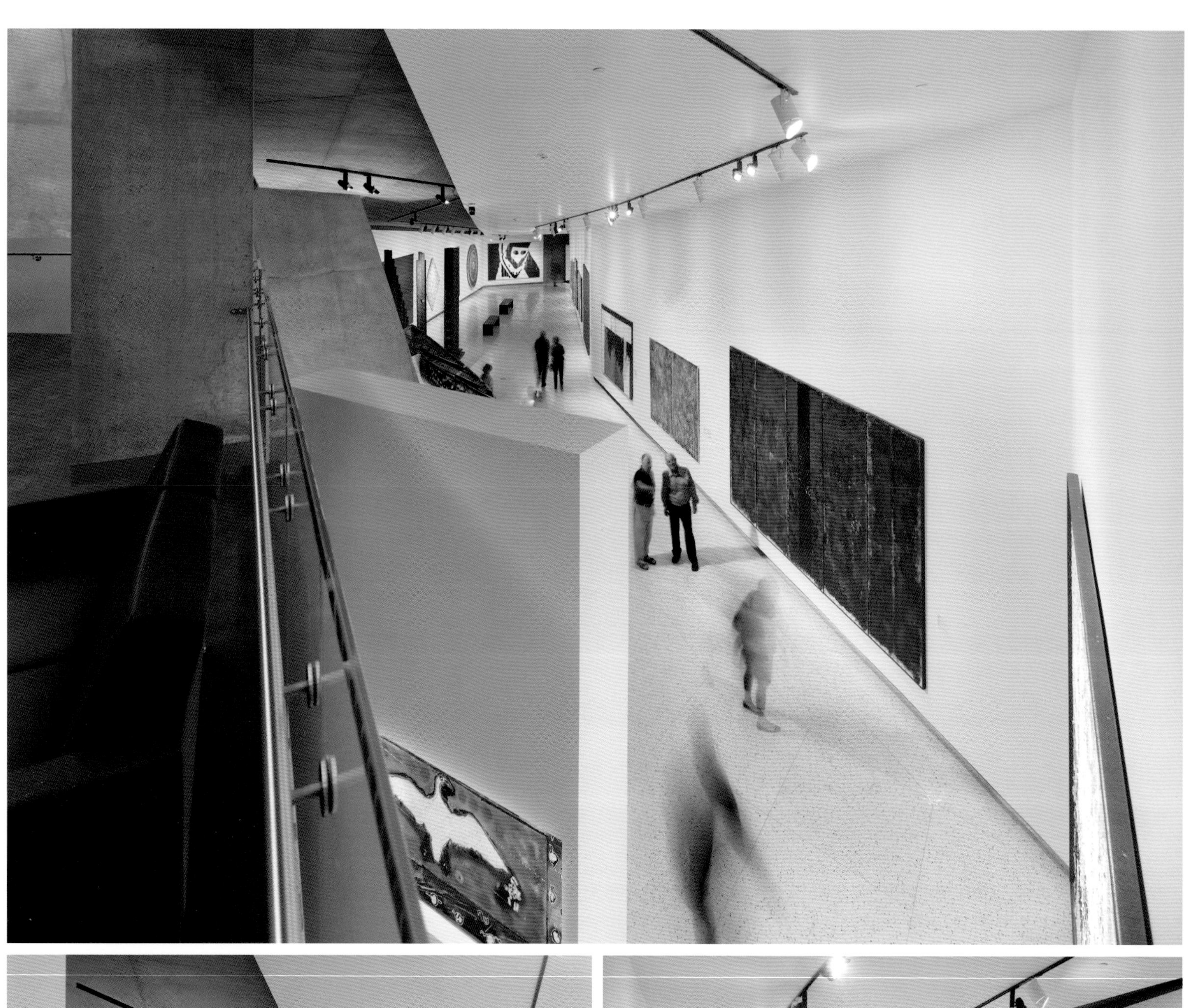

条形建筑

办公建筑

Central Office of FEDA Confederation of Employers of Albacete

FEDA 办公楼

Design Company / 设计事务所:
Cor & Asociados
Project Architect / 设计师:
Miguel Rodenas, Jesús Olivares
Location / 地点:
Spain （西班牙）
Size / 面积:
4375 m^2
Photography / 摄影:
David Frutos

视觉亮点

建筑外墙应用特殊材料，这种材料既保护了建筑外墙，又使立面更加丰满。

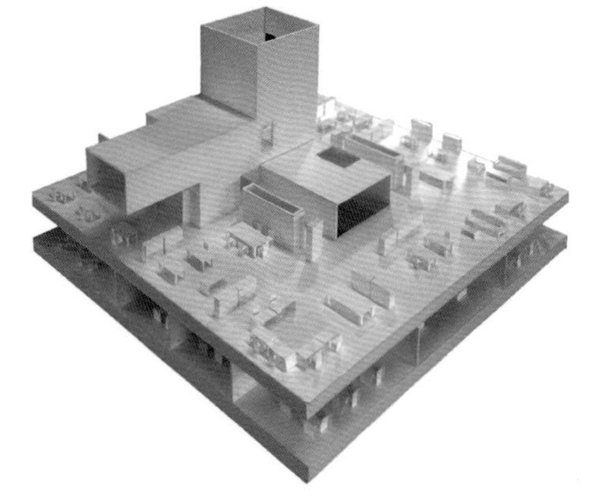

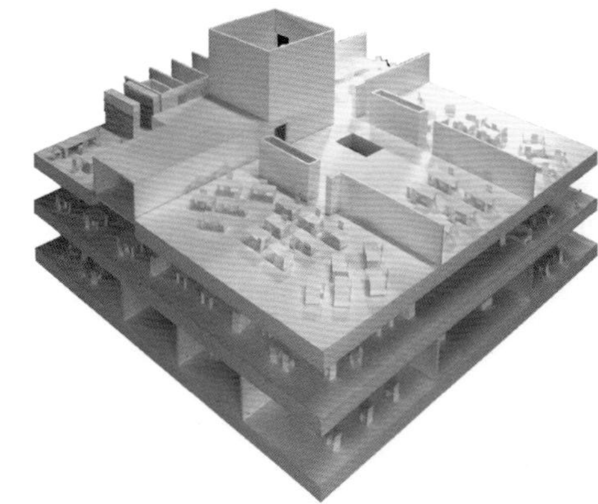

feda

feda

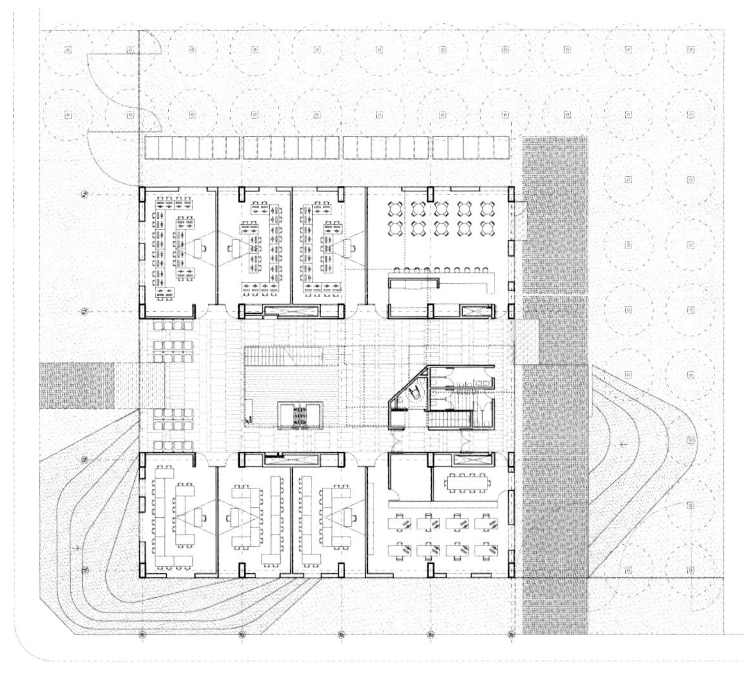

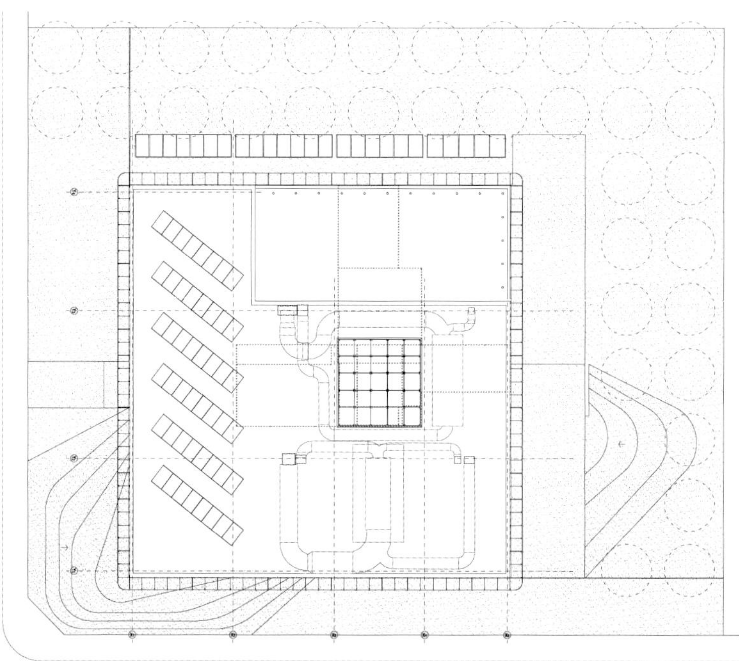

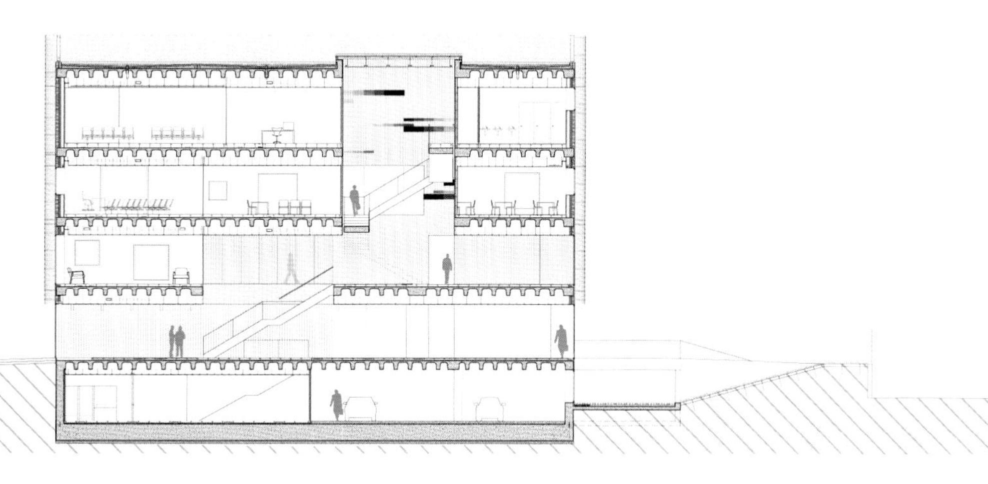

We have designed this project from the idea of 'diffuse limits' and 'blur' architecture. Our intention was to cover the volume of the building with a veil capable of blurring it and making it change. We wanted the building to react to the variations of weather and the movement of users with different levels of brightness and textures.
Looking at it from outside to inside, the skin would feel 'fleshy', full of shades and thick. And at the same time it would appear as a distant and undefined object, so that the observer doesn't have a stable reference, and could not keep a static link to the building and remember only an image. On the contrary the building would respond to the user in movement generating different glances and changing perceptions.
In the opposite view, this second skin had to be perceived as a space with constant shape and without scale changes. Likely, the inner façade with the windows is the one able to defragment the building because the windows are very large compared to the human scale.

设计师打造该项目的最初理念是“消除限制性条件”，并对建筑进行“模糊化处理”。因此，设计师用类似帷幕的外框架将整个建筑结构遮盖起来。经过打造，建筑在不同的光照条件下和使用者所处的位置不同时会展现出不同的特点。
当人们从建筑外部向建筑内部看过去时，建筑外立面显得相当“丰满”，具有丰富的光影效果。同时，建筑又像是一个遥远的、不能明确定义的物体，观察者并没有一个固定的参照物，头脑中只能有一个大致的景象。另一方面，由于观察者所处角度和视线的不同，建筑也会呈现不同的形象。
建筑的第二重立面是那些拥有连续式外观的空间，建筑看起来并不具有规模上的变化。设置了窗户的第二重立面进一步表现了建筑的空间特色。

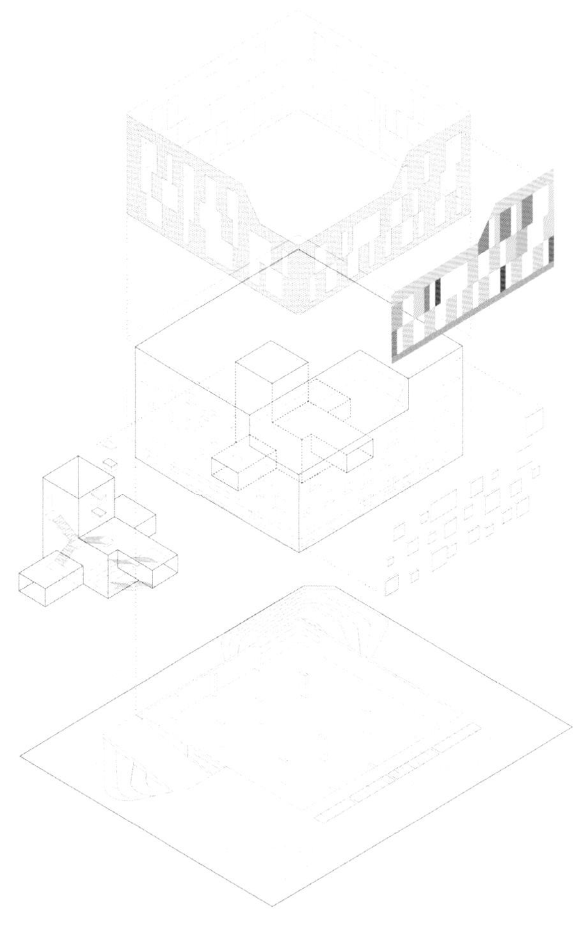

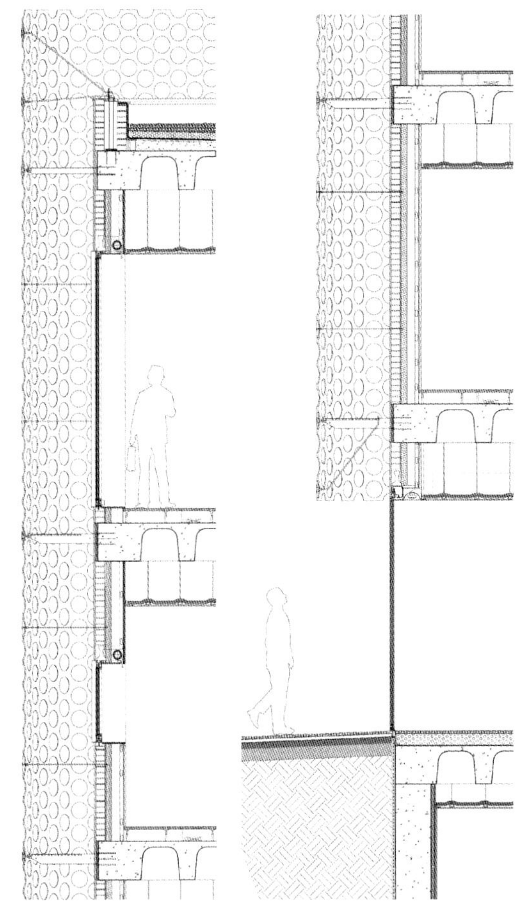

feda

条形建筑

办公建筑

Prosecutor's Office

检察官办公楼

Design Company / 设计事务所:
Architects of Invention
Design Team / 设计团队:
Niko Japaridze, Gogiko Sakvarelidze, Ivane Ksnelashvili, Dato Canava, Eka Kankava, Viliana Guliashvili, Nika Maisuradze, Elisso Sulakauri, David Dolidze, Soso Eliava, Eka Rekhviashvili, PM Devi Kituashvili
Location / 地点:
Georgia（格鲁吉亚）
Area / 面积:
site 3 000 m^2, building 752.78 m^2, total floor 1 500 m^2
Photography / 摄影:
Nakanimamasakhlisi Photo Lab

视觉亮点

建筑的外观时尚且具有动感。正立面由多个长方形体块构成，这些长方形体块凸出并相互错开，形成了一个个独立的空间。

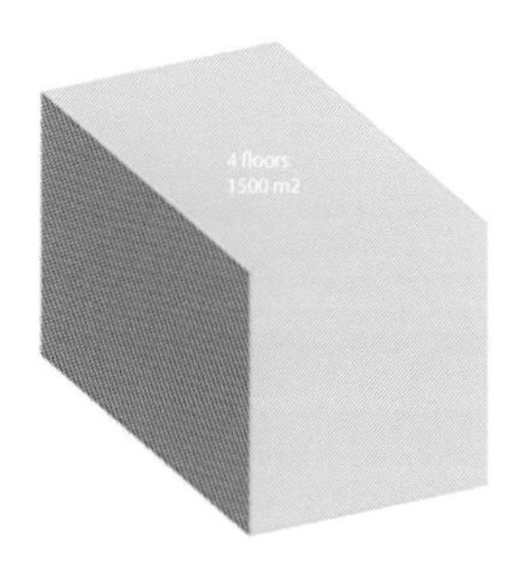

1. Required Programme

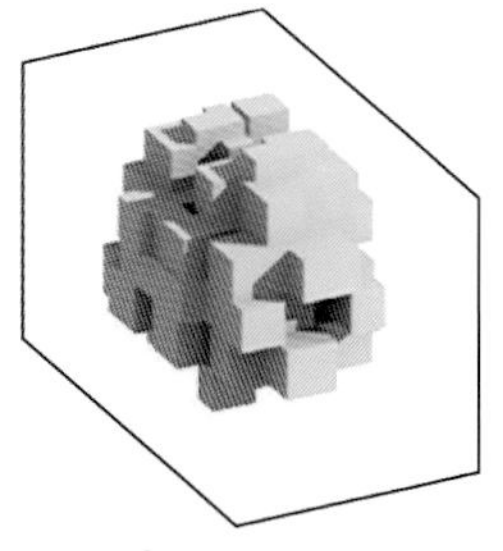
2.Dividing programme according functions

3. Forming Atriums in Volume

4. Final solution

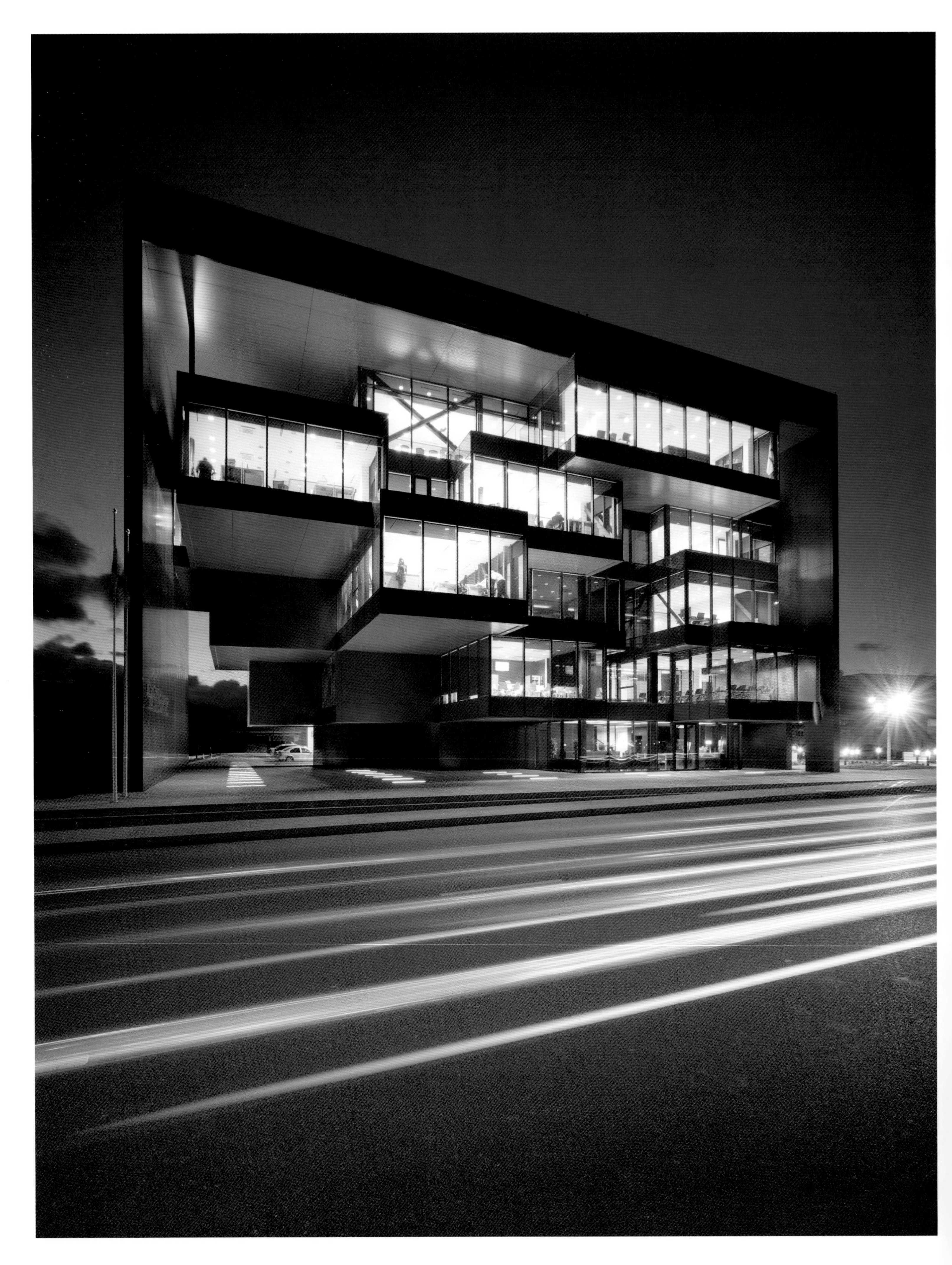

The building consists of a stalwart, stark black frame that has luminous, delicate volumes suspended within it. 70% of the building is freestanding away from the sides and the ground creating an impression of 'floating' forms. The volumes make way for a passageway to one side of the building, creating a physical transparency of the building that is alluded to throughout the design. The suspended volumes are staggered so that they create terraces and balconies.

The interior intelligently uses glass to create a feeling of transparency and light, in contrast to the dark impenetrable framework. The ground floor serves as the main access area with security check in and lobby. The top floor is dedicated to the Prosecutor's office, a meeting room, canteen and veranda for use by staff. A sky park, or roof garden, is used for conferences and celebrations. In contrast to the openness of the front facing facade, the back of the building is smooth and functional with few windows.

建筑为黑色的大理石框架，70% 的建筑空间是独立式悬挑结构，建筑的一侧是中空结构。建筑的正立面应用了大面积的玻璃材料，为室内提供了较好的采光条件。悬挑式结构互相错开，以便提供更多阳台空间。

建筑内部也使用了玻璃材料，形成了一个明亮、通透的办公空间，与密不透风的空间形成对比。一层设置了安检区域，顶层设置了检察官的办公室、会议室、食堂、供全体职员休息的露台。屋顶花园为举办活动提供了场所。与正立面开放的状态相比，建筑背面比较平整，只是设置一些具有通风功能的窗户。

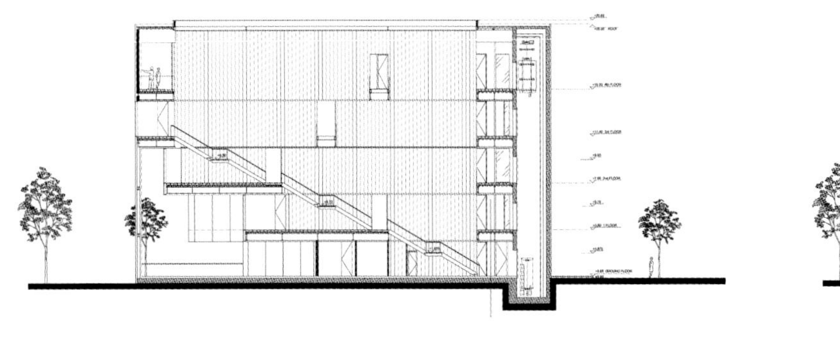

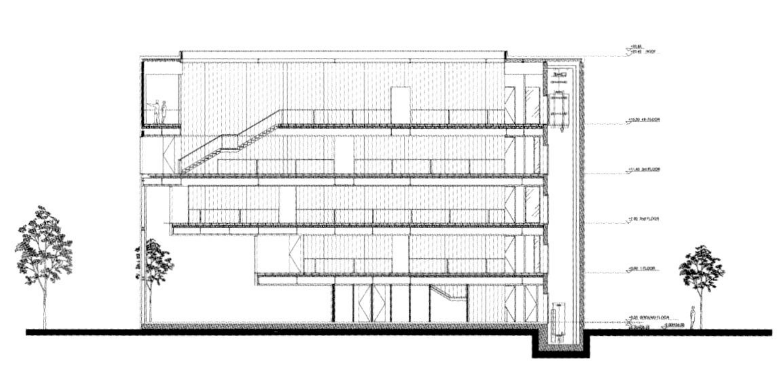

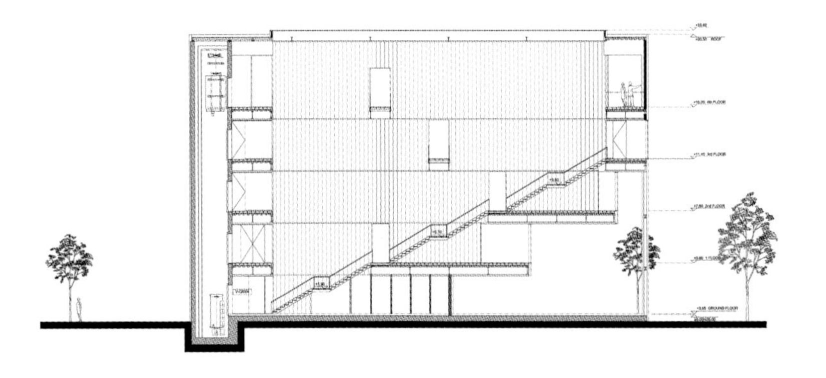

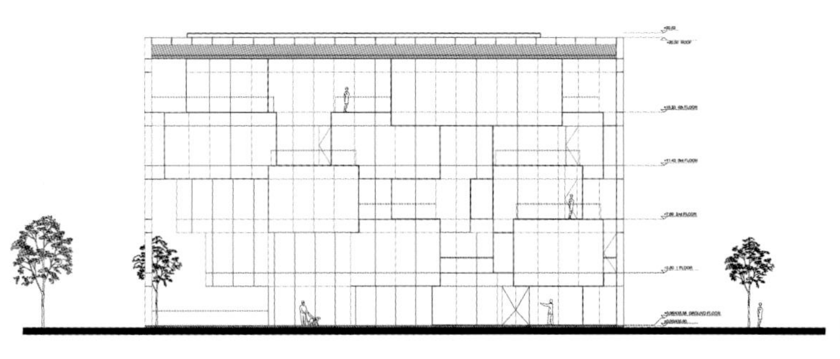

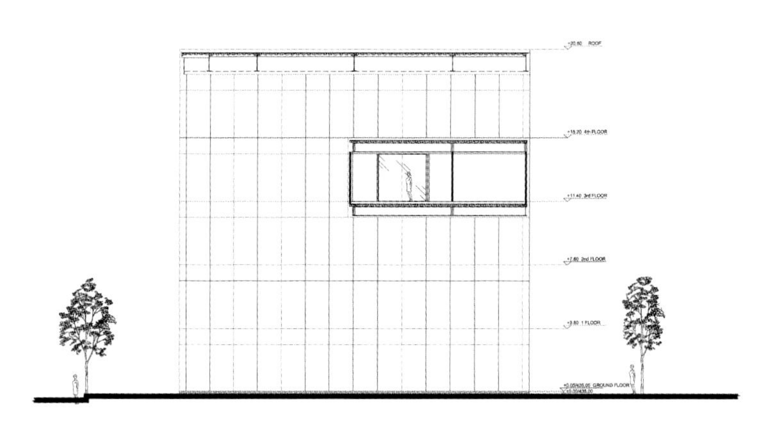

Office

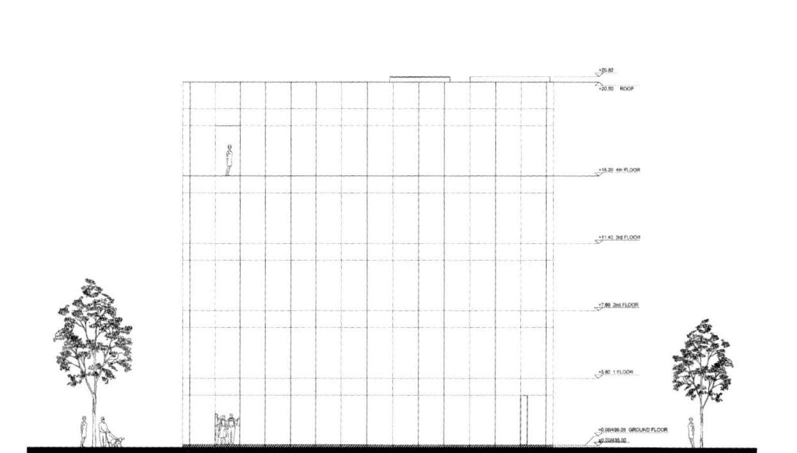

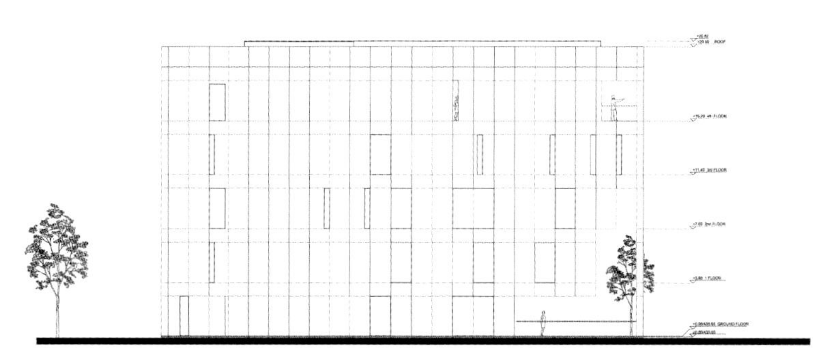

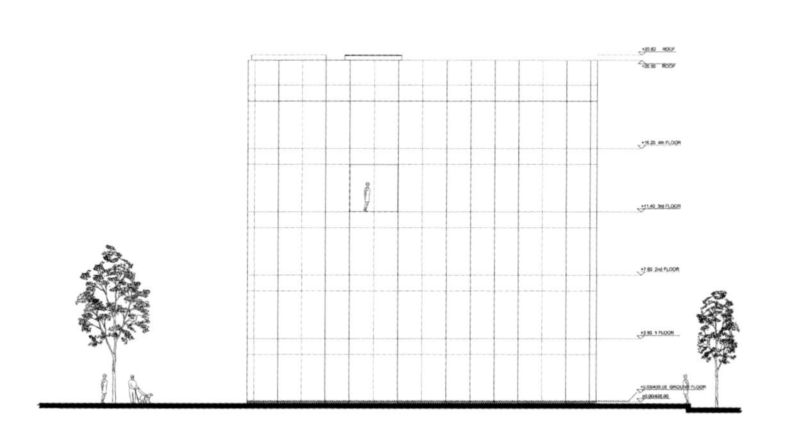

TELEPHONE

条形建筑

办公建筑

PITA and TECNOVA Headquarters

PITA and TECNOVA 办公总部

Design Company / 设计事务所:
Ferrer Arquitectos
Project Architect / 设计师:
José Ángel Ferrer
Location / 地点:
Spain （西班牙）
Area / 面积:
Pita Building 15 028.58 m^2, Tecnova Building 1 605.60 m^2, Both Buildings 16 634.18 m^2
Photography / 摄影:
David Frutos

视觉亮点

建筑外立面极具线性特征，建筑内部拥有类似庭院的完美空间。

PITÁGORAS

PITÁGORAS

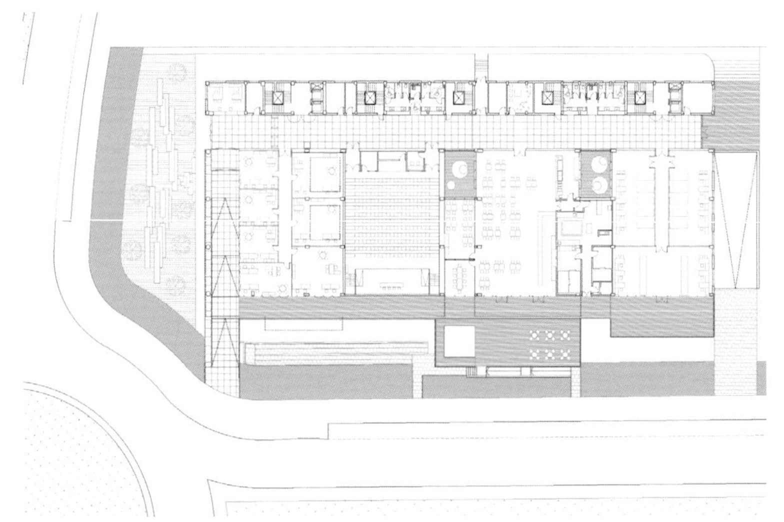

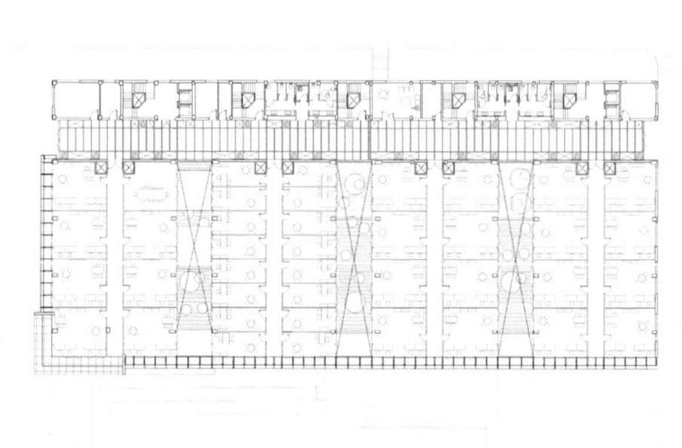

These buildings are laid out according to a strip arrangement, creating an ordered and functional design which sets out clear entrances and highly versatile, multi-purpose spaces. There is a main communications strip, which is the backbone of the design, created as a glazed area with climbing vegetation and scattered courtyards that provide interesting sequences for the visitor/user. Each floor contains a meeting room that protrudes out over the landscape, offering views of the bay and the Gata Cape Natural Park. Both buildings exceed the minimum environmental standards and have been designed to meet energy classification A, as well as the American LEED certification and the European Green Building Certification.

建筑布局合理，建立了一个有序和功能齐全的设计，并设置了明确的出入口。建筑为横向布局。建筑的有些空间栽种了攀缘植物，这些植物为使用者创建了一个有序的类似庭院的空间。建筑每层都有一间会议室，这些会议室拥有观赏海湾和公园美景的绝佳视野。建筑生态环保，获得了美国 LEED 认证和欧洲的绿色建筑认证。

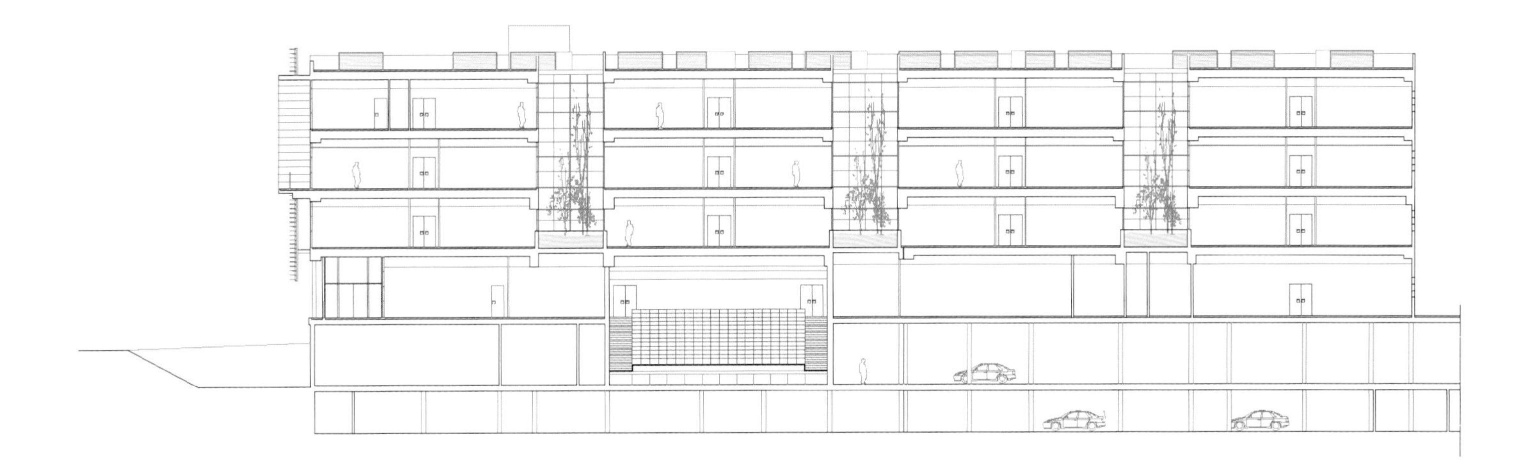

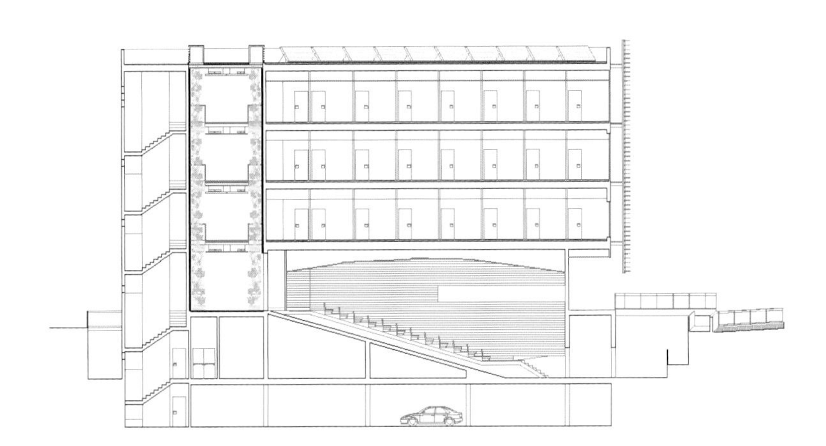

ARCHIVO 1

Gazoline Petrol Station

加油站

Design Company / 设计事务所:
Damilanostudio Architects
Project Architect / 设计师:
Duilio Damilano
Location / 地点:
Italy （意大利）
Area / 面积:
3 800 m²
Photography / 摄影:
Andrea Martiradonna

视觉亮点

建筑立面简洁、素雅，为人们提供了一个舒适的休憩环境。

The design of a service station is a strong reference to the idea of travel, short or long-distance routes interrupted only by a few stops and then back on the road.A break for refueling, or just to stretch a bit 'legs before continuing his journey.With the same continuity, the service station is separated from the asphalt like a ribbon of road with the engine and wrapping around itself, creating a temporary volume to accommodate the traveler.

The architecture of the service station, as usually conceived as a mere support function, thus acquires a shape. The architecture, static by definition, becomes closely linked to the concept of continuous flow that envelops and becomes the urban landscape without interruption.The shell reinforced concrete, cast in special molds fluidized is closed by glass walls.Are distributed within the office manager and a self-service, separate bathrooms from the block in central position. On the rear elevation red steel block is detached from the body and a wolf howling, illuminated at night, draws attention to the urgent needs.

本案的设计与旅游的概念紧密相连。旅游时，人们可以停下来加点油，还可以在继续旅程之前伸伸腿。服务站与公路有一点距离，创造了一个临时的供游人休憩的地方。

服务站通常被认为是只具有服务功能的建筑，这影响了它的外观设计。该建筑被设计成了静态的形式，成为城市景观的延续，并与其紧密联系在一起。

用特殊模具现浇的钢筋混凝土结构上设置了大面积的玻璃幕墙。办公空间、自助服务空间、浴室围绕着中央区有秩序地分布。一处红色的结构与主体结构分离。一个在夜晚被照亮的大灰狼图案吸引着人们进入洗手间。

GAZOLINE

GAZOLINE ST

条形建筑

INGFAH Restaurant

INGFAH餐厅

Design Company / 设计事务所:
Integrated Field
Location / 地点:
Thailand （泰国）
Area / 面积:
site 2 040 m², gross 430 m²
Photography / 摄影:
IF & Wison Tungthunya

视觉亮点

方形的框架结构，就像一个个灯笼，在夜晚，点亮了整个用餐区域，为餐厅增添了无限的乐趣和魅力。

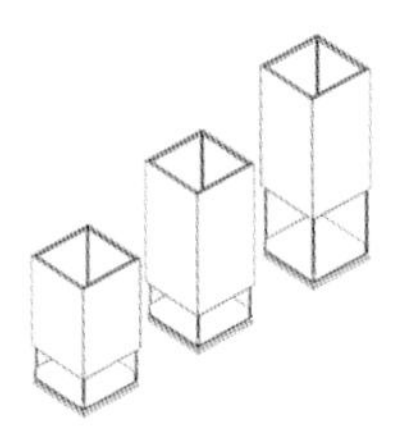

Seating unit : 4m / 5m / 6m

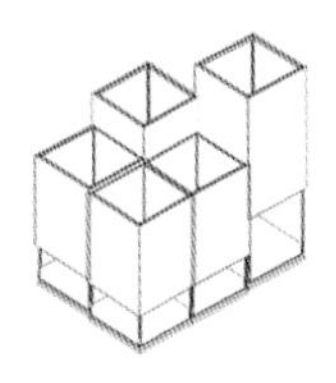

Seating units in group

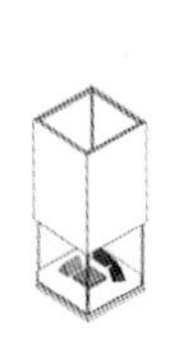

Bed unit

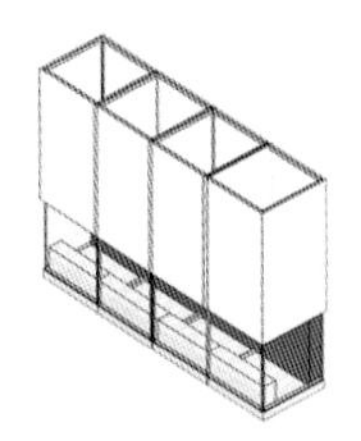

Bar unit

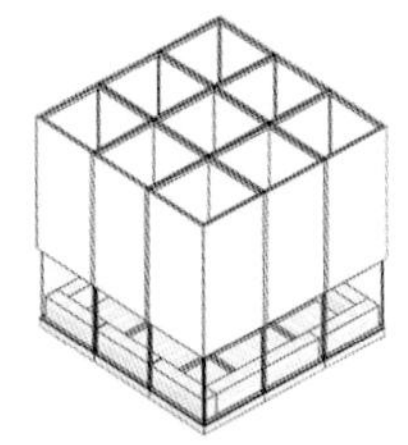

Grill bar unit

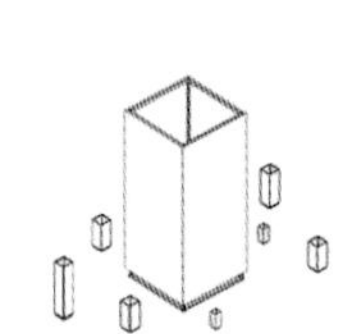

Tree / Lighting unit

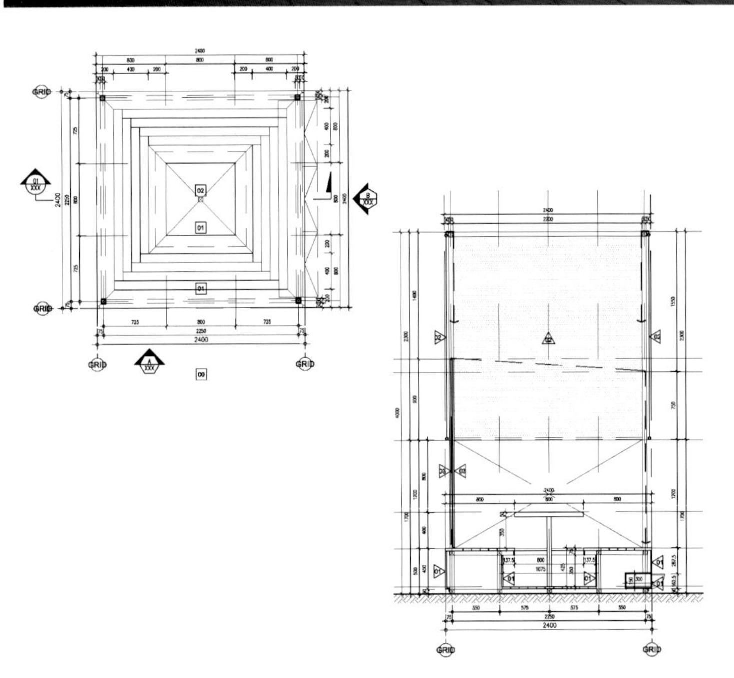

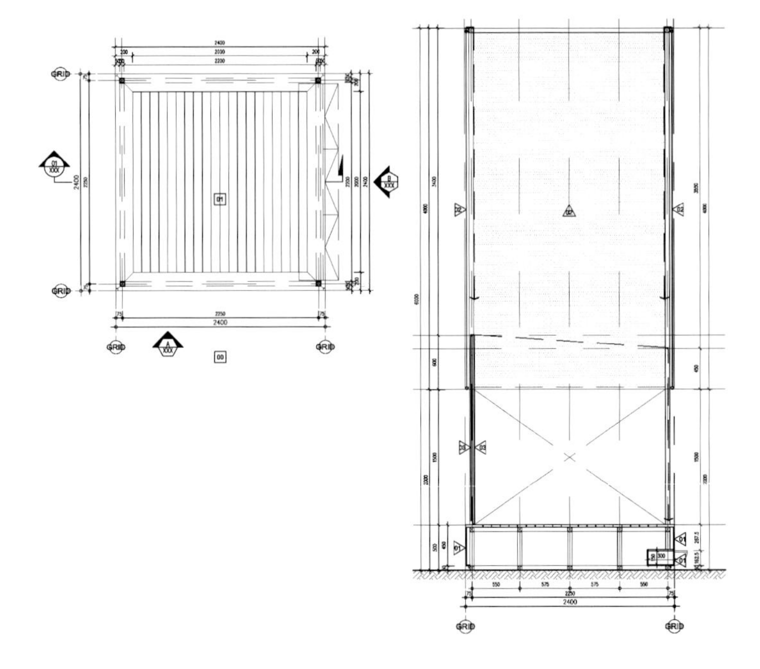

First time when we arrived the location, the site of this restaurant is surrounded by buildings and has no sea view. The most effective potentials left for this site was the lawn and the sky.

Therefore, IF proposed the concept of dining in the new way, sitting / lying down (which adapted and redefined from the past Thai dining behavior), and "Frame" the sky view to capture and make the beauty of the sky even more meaningful. Then, we had considered about the constraint of the limited construction time and the owner's requirement that want this project to be a new destination for people, and also, raise the standard of the restaurant level around Khoalak area, which could be like lightening this area with flying lanterns spread out in the sky at night.

The conceptual idea came to be the restaurant that contain "many of small units" which can be constructed in the shorter period than one big unit, and easier to be re-arranged in the future. Furthermore, IF had proposed the new style of dining and the food itself. The main structure for each unit was designed to be a "light geometric structure", the very slim steel skeleton structure. It was meant to blend in with the sky and let only the fabric stretched on it exist to the people's sight. This fabric would be the "frame that capture the sky" and "the lantern that light up that area".

当设计师第一次来到该项目场地时，看到的是这样的景象：饭店周边被其他建筑所环绕，望向大海的视线被遮蔽起来。设计师决定充分利用建筑周边的草坪和美景。

设计事务所构思出新的方式来打造用餐区，用餐区可供人们坐着或者躺着来享受美食（对泰国人以前的用餐方式进行了一些改良）。设计师设定了欣赏天空的视野，使人们能够欣赏美丽的天空。设计师在设计过程中要考虑到有限的建造工期及客户的要求——让这家餐厅成为人们用餐的新的目的地，从而提升该地区的就餐环境。设计完成后，餐厅要像散布在苍穹中的一个个灯笼，点亮整个夜空。

该饭店的主要设计理念是设置许多小型单元。比起大型结构而言，这样的设计不仅确保了整个项目可以在短期内完成，也便于在将来对整个空间进行重新配置。每个单元的主结构采用的是轻钢型框架结构。设计师的主要设计意图是使结构单元与天空融为一体，仅让每个框架构造的上部出现在人们的视野范围内，而这部分构造将是“捕获天空的框架”和“点亮的灯笼”。

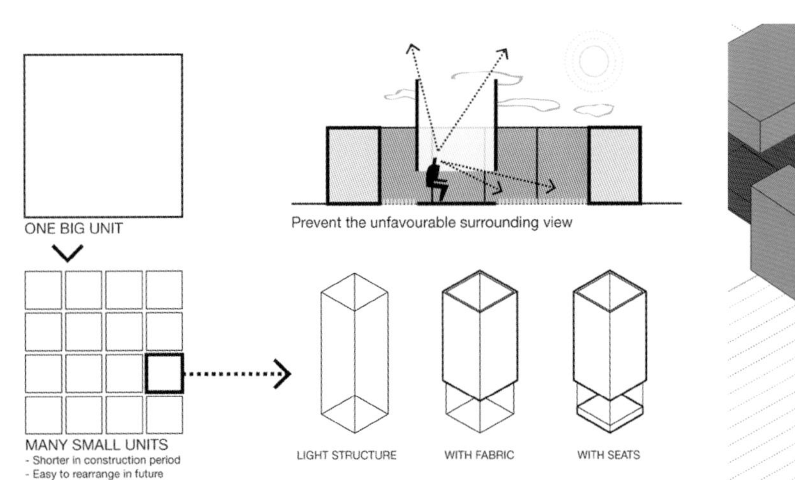
ONE BIG UNIT
MANY SMALL UNITS
- Shorter in construction period
- Easy to rearrange in future
Prevent the unfavourable surrounding view
LIGHT STRUCTURE
WITH FABRIC
WITH SEATS

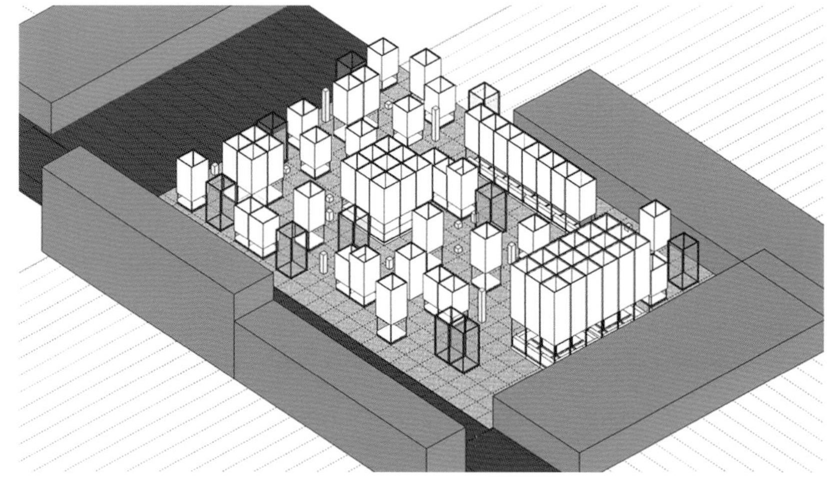

L.S.G. Head Office Building

L.S.G. 总部办公楼

Design Company / 设计事务所:
Urban Office
Project Architec / 设计师:
Lilian & Iana Captari
Location / 地点:
Romania （罗马尼亚）
Area / 面积:
2 200 m^2
Photography / 摄影:
Arthur Tintu

视觉亮点

建筑立面上凸出的类似箱体的结构，不仅使建筑立面更为丰富，还在建筑本身与周边住宅区之间建立了平稳的过渡。

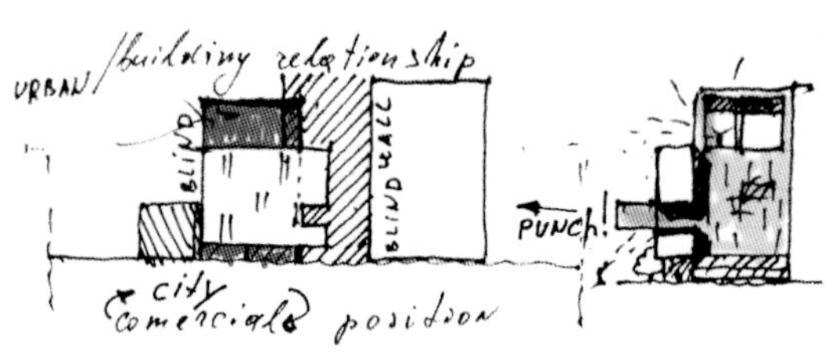

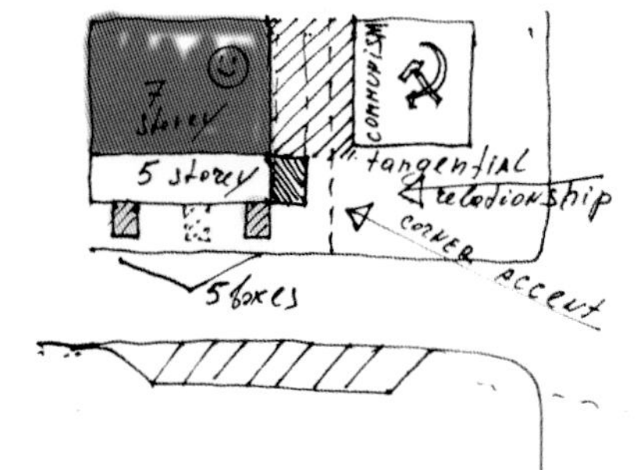

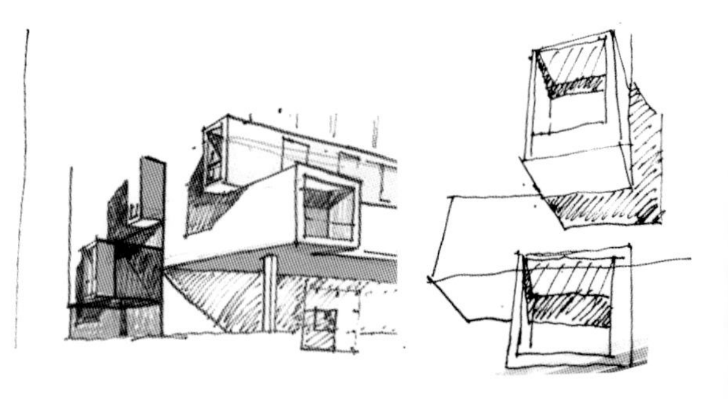

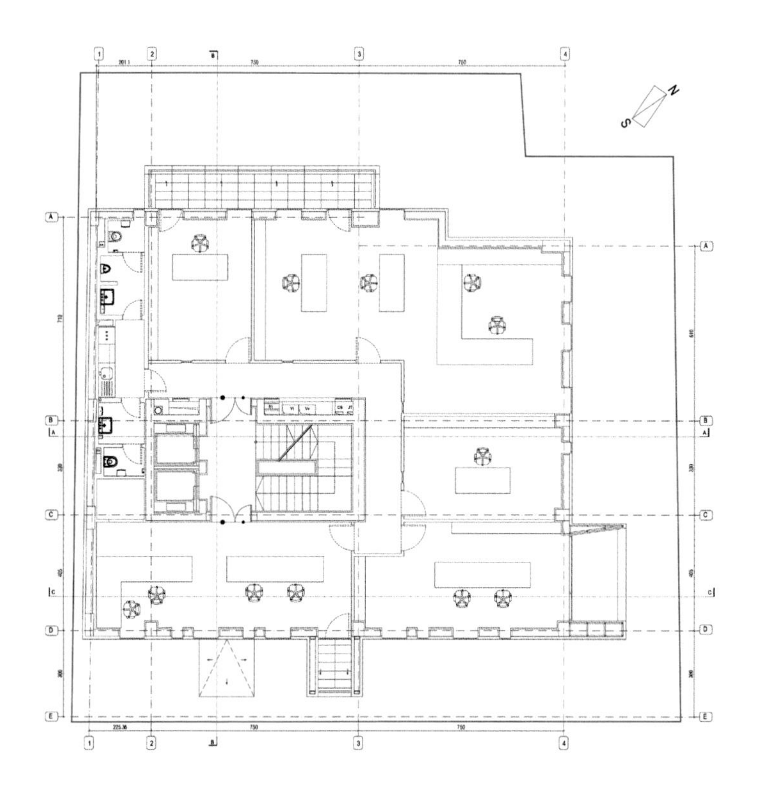

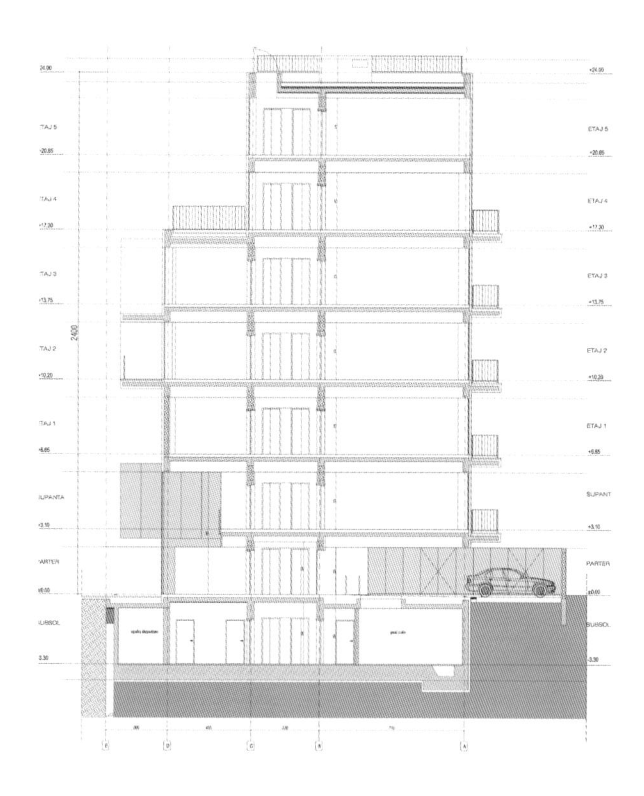

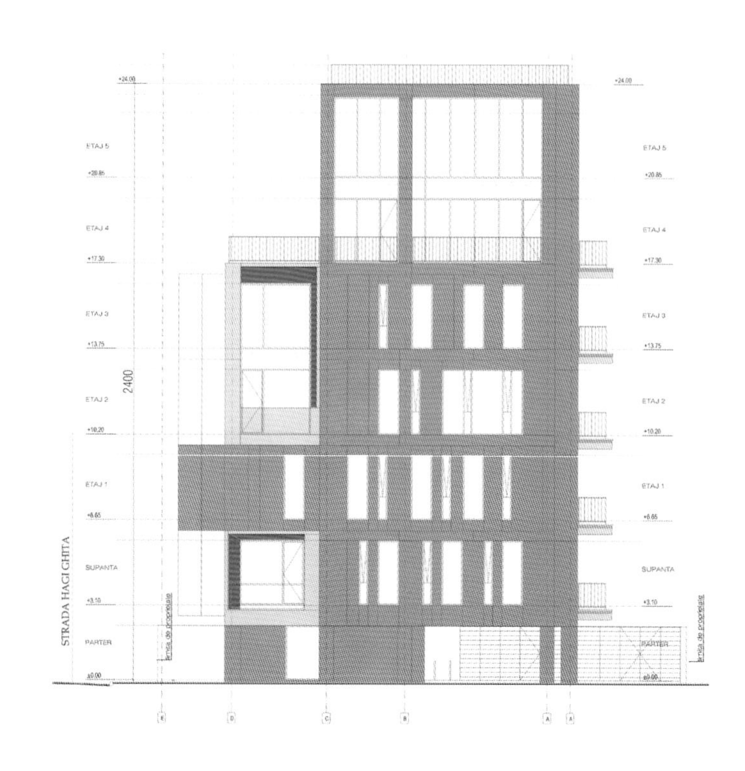

Compound geometry of the building is defined by boxes which "bombard" or dot the façade, which confer a certain degree of urban porosity, meant to reduce the massiveness of the building, in order to offer a gradual transition to the surrounding residential area. Having such powerful vibrations on the façade, we decided to attentively and minutely divide the rest of the façade, using vertical panels with minimal joints. Windows are placed in the same line with the finishing, so as to achieve that flatness specific to Northern architecture.

Ground floor boasts an above ground parking lot, a patio and enjoys free access from all directions.

By using laborious details we tried to achieve an elegant finishing to suit cantilevered volumes, which play multiple roles in this composition… they serve as daytime smoking areas from a functional point of view and overnight they turn into night lighting elements, integrated into the building computer (BMS).

建筑的外立面上设置了许多凸出的类似于箱体的结构，这种结构展示出了多孔状的城市空间结构，使建筑显得不那么厚重，在建筑本身与周边住宅区之间建立了平稳的过渡。基于如此富有特色的立面设计，设计师决定遵照此原则对立面其他部分进行精细划分，主要通过设置垂直板材来实现空间的划分，并设置尽可能少的接合点。窗户沿同一条直线布置，以打造出罗马尼亚北部建筑所特有的整齐的特点。

建筑一层设有一处地上停车场和一处露台，人们从各个方向上均可自由进入这处空间。

通过对一些结构细节仔细雕琢，设计师打造出优雅的建筑形态，使悬臂式结构与其他建筑结构融为一体。而悬臂式结构对于整座建筑来说也具有多重功能……在白天，从空间功能的角度来看，这些空间可用作吸烟区；当夜幕降临时，它们又变成了发光体，并纳入整座建筑的照明系统中。

ABC Museum

ABC博物馆

Design Company / 设计事务所:
Aranguren & Gallegos Arquitectos
Project Architect / 设计师:
Maria José Aranguren Lopez, José Gonzalez Gallegos
Location / 地点:
Spain（西班牙）
Area / 面积:
3412.5 m²
Photography / 摄影:
Jesus Granada

视觉亮点

三角形构成长方形里面的主要元素，建筑立面干净、简洁，极具现代时尚感。

"Architecture is a mixture of NOSTALGIA and ANTICIPATION, the co-existence of history and vanguard".(JEAN BAUDRILLARD)
The current building provides access from two streets, connecting these with an internal patio.One of these accesses is currently offered under a one-floor high, longitudinal building body closing the internal patio towards the street.The main doors for access to the new ABC Centre are considered on this front. The aforementioned longitudinal body is restructured for this, as a large translucent glass "beam" that works as the lintel of a gap for passage to the internal patio. The cafeteria will be housed inside this, with the basement floor of the new centre receiving light through a glass floor for access to the patio.

该项目的设计理念可以用一位知名建筑师的话来概括："好的建筑设计是将怀旧情怀与对未来期望结合在一起，将历史和现代同时呈现在人们眼前。"现有的这座博物馆建筑在两条街道上都设有入口通道。一处一层楼高的纵向建筑结构将博物馆的内庭与街道分隔开来。而前面提到的两条入口通道中的一条就位于这处纵向结构的下方。进入 ABC 博物馆的主要入口大门也位于建筑的这一侧。设计师对纵向建筑结构进行了重新打造，营造出大型的通透式玻璃"梁架"结构。博物馆的餐厅也位于这一侧。通向内庭的地板为通透式玻璃结构，这使得自然光可以射入地下室，节约了大量电能。

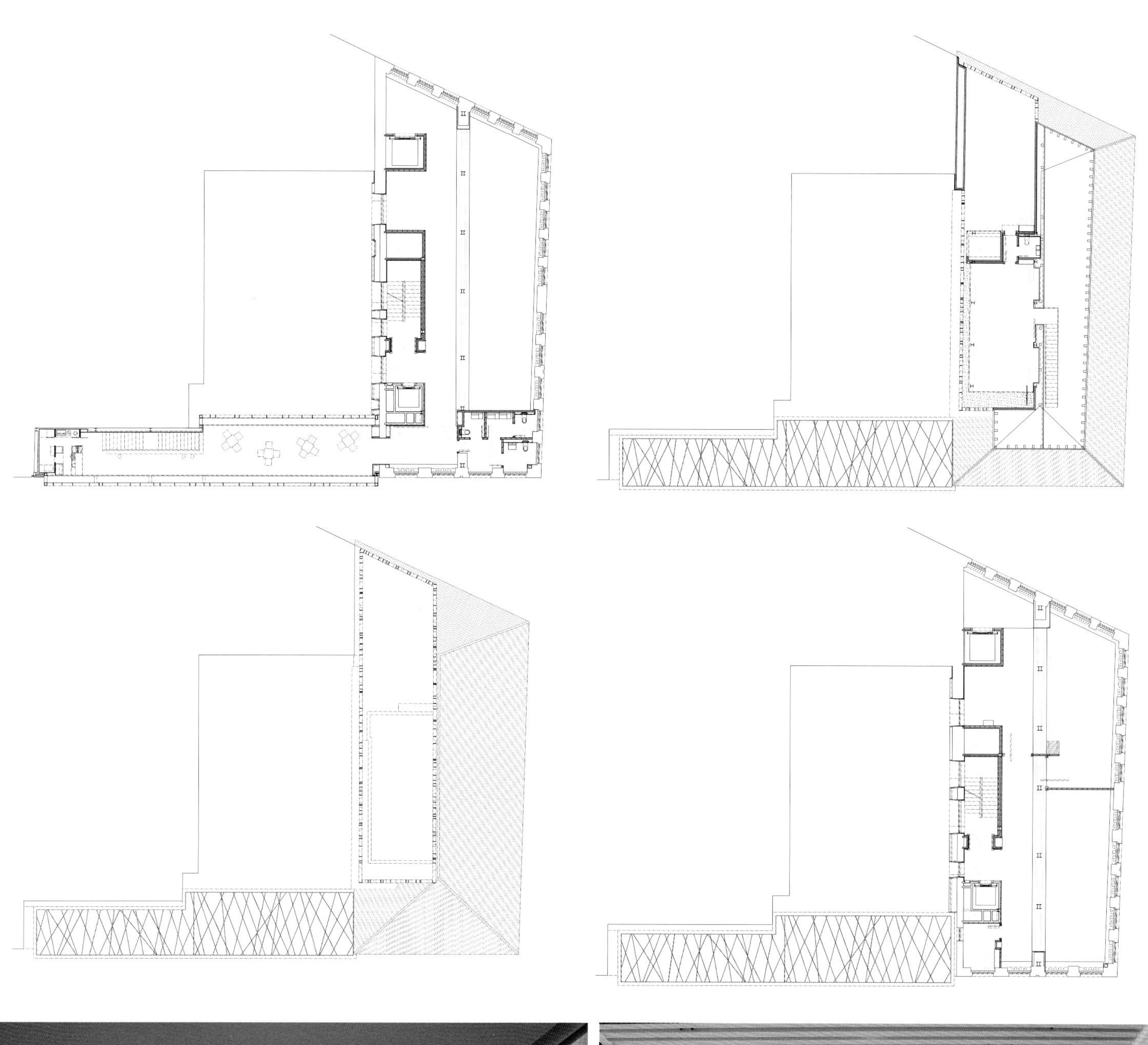

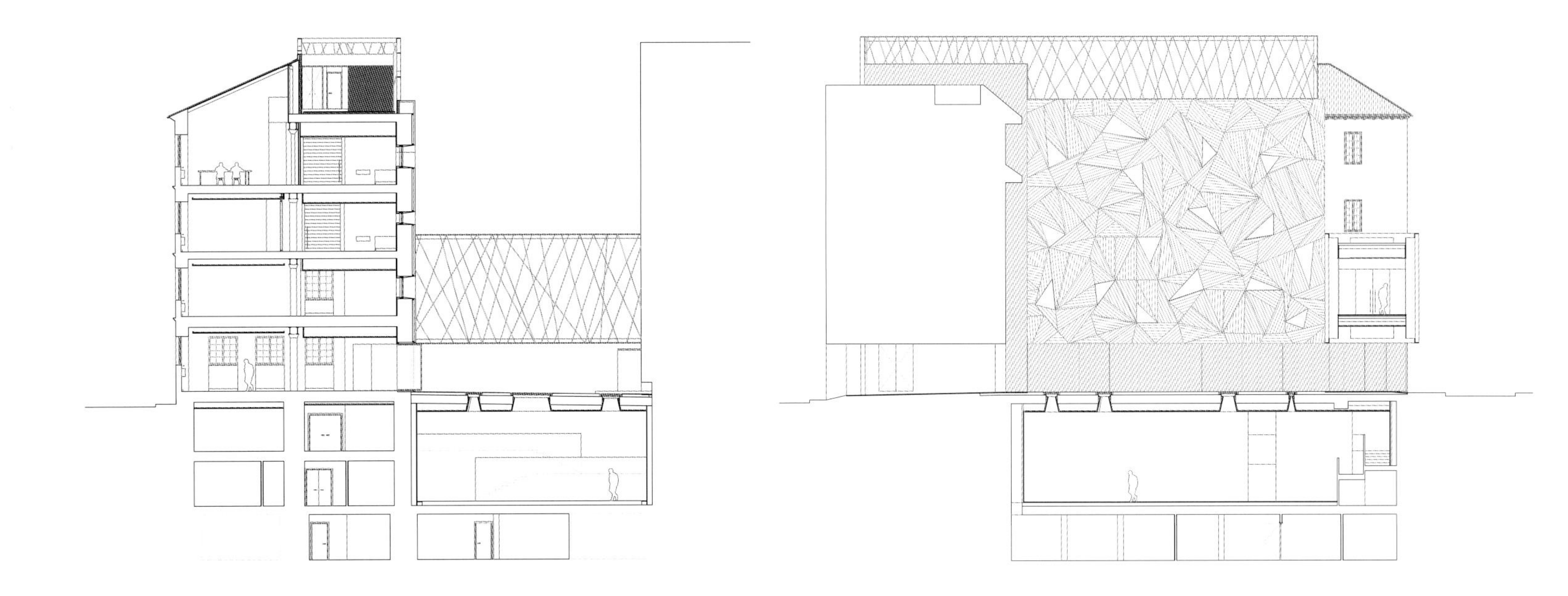

条形建筑

办公建筑

Westerlaan Tower

韦斯特兰办公楼

Design Company / 设计事务所:
Ector Hoogstad Architecten
Project Architect / 设计师:
Jan Hoogstad, Joost Ector , Max Pape, Gijs Weijnen, Fleur Doelman
Location / 地点:
The Netherlands（荷兰）
Area / 面积:
offices 8 500 m^2 , residencies 13 700 m^2
Photography / 摄影:
Petra Appelhof, Ossip van Duivenbode, Maarsen groep, Marcel van Kerckhoven

视觉亮点

建筑外观清新、简洁，黑白色彩的搭配使用使建筑外墙更显干净、纯粹。

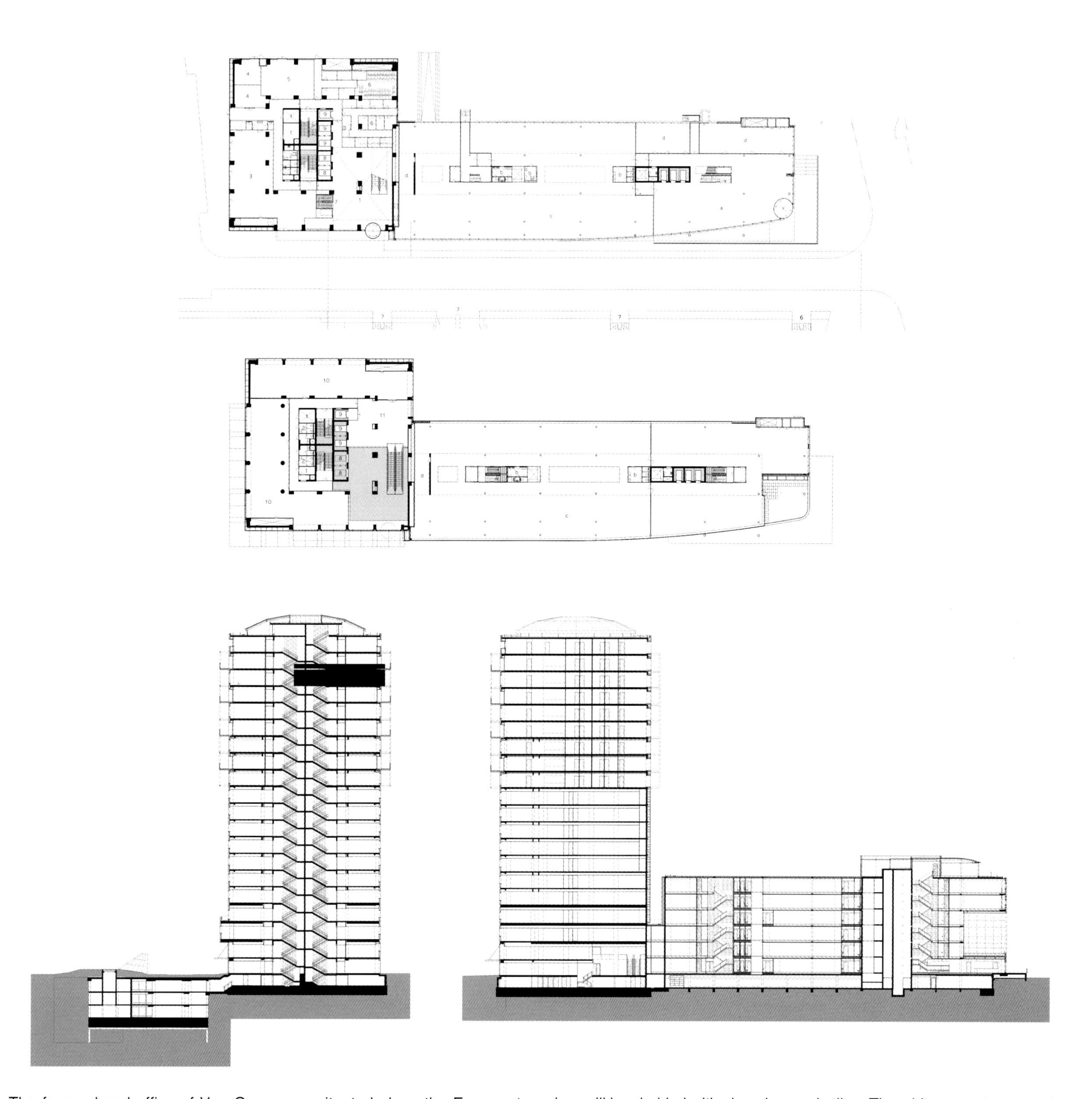

The former head office of Van Ommeren, situated along the Euromast Park and the Maas river, was made up of low-rise and a tower. Through clever design, we efficiently created extra space within the existing volumes of low-rise and the tower. As a result, the new Vopak head office could be accommodated in its entirety in the completely revamped low-rise, leaving the tower free for other use.

We utilised the constructional overcapacity of the office tower to create three extra floors. It was possible to demolish the top two floors and replace them with five new ones of a lighter construction, on condition that dwellings were included.

A new facade will be attached to the existing structure. Within a new steel framework, spacious balconies with glass parapets and windbreaks will be added. The new facade has windows and large sliding French windows within the framework of the balconies. The tower, like the low-rise, will be cladded with glazed ceramic tiles. The white accents connect the tower with the low-rise, so that they form a whole visually.

该建筑曾是一家公司的总部办公楼，建筑附近有一座公园。整个项目由一座低层建筑结构和一座高层建筑结构组合而成。设计师通过巧妙的设计，在低层和高层结构的原有空间内创造了更多的空间。重新打造的低层建筑可以容纳集团总部的所有办公空间，这样高层建筑就可以设置其他使用功能的空间了。

原有顶部的办公楼空间没有被充分利用起来，设计师将顶部两层拆除，并使用轻型结构重新打造了五层新的结构，其中一部分被设计为高级公寓。

设计师给原有的建筑结构新加了一处立面。在新建的钢结构框架内设置了开阔的阳台。新建的阳台拥有玻璃材料制成的栏杆和防风墙。阳台的框架结构内还设置了大型的滑动落地窗。与低层建筑一样，高层建筑外围也覆以釉面瓷砖。高、低建筑外墙使用同样的黑白相间的色彩，使整体建筑风格统一。

Théâtre de Quat'sous

德奎特苏斯剧院

Design Company / 设计事务所:
Les architectes FAB
Location / 地点:
Canada （加拿大）
Photography / 摄影:
Steve Montpetit

视觉亮点

建筑外墙使用了多种不同颜色、不同质地的材料，使原本造型简洁的建筑丰富多彩起来。

Théâtre de
Quat'Sous
ABONNEZ-VOUS

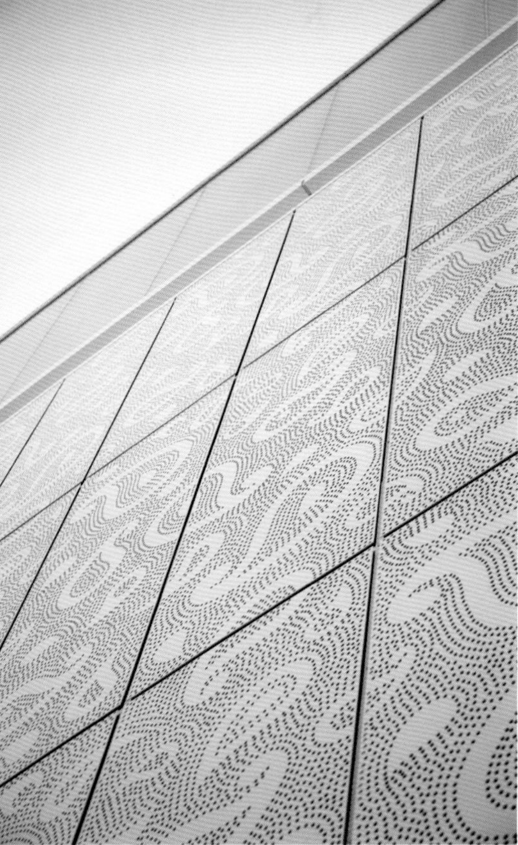

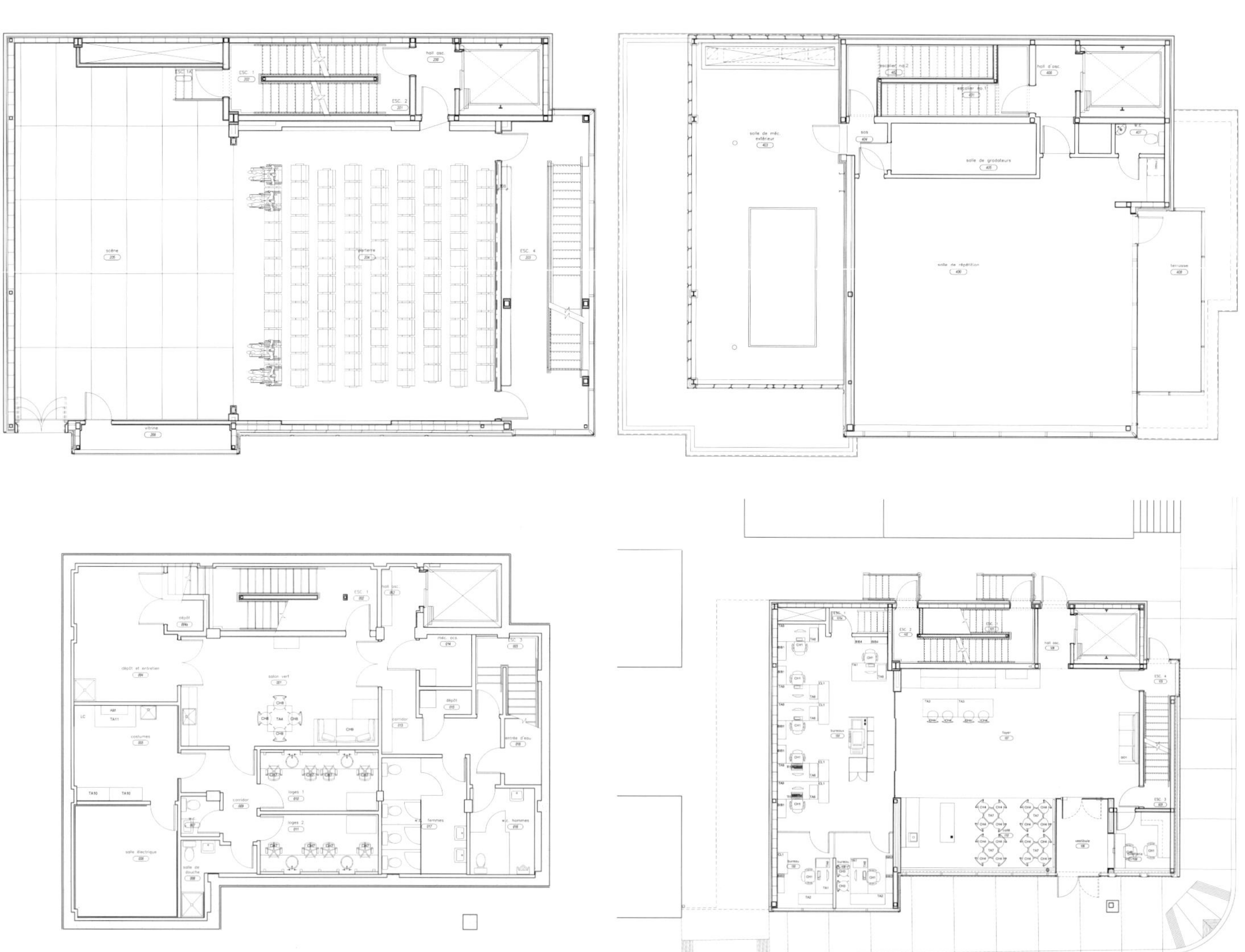

In reconstructing the Théâtre de Quat'sous we were specifically asked to incorporate whatever we could from the original building to help the new one evoke those memories. We chose to sample textures, images, colors and materials from a cultural inventory of the theater and mapped them on the assemblage of required volumes (stage, house, foyer, crossover, control booth and rehearsal). Recycling on site stones, slate, wood, bricks, marble and furniture becomes part of a strategy of cultural sustainability. New materials include silkscreened glass, black brick and perforated aluminum that contribute to make Théâtre de Quat'sous a ghostly figure accumulating memories.

在该剧院重建过程中，业主对项目团队提出了一个特别的期望：从原有老建筑中汲取灵感，使新的建筑能重现那些被人忘却的记忆。设计师从剧院的文化角度入手进行研究，对剧院的质地、形式、色彩和材料进行精心设计，并将其体现在需要打造的建筑结构中（比如舞台、门厅、控制室、排练厅等）。文化可持续性的设计策略还体现在对现场的石材、板岩、木材、砖块、大理石、家具等的重复利用上。此外，该建筑还采用了一些诸如丝网玻璃、黑砖、多孔铝的新型材料。

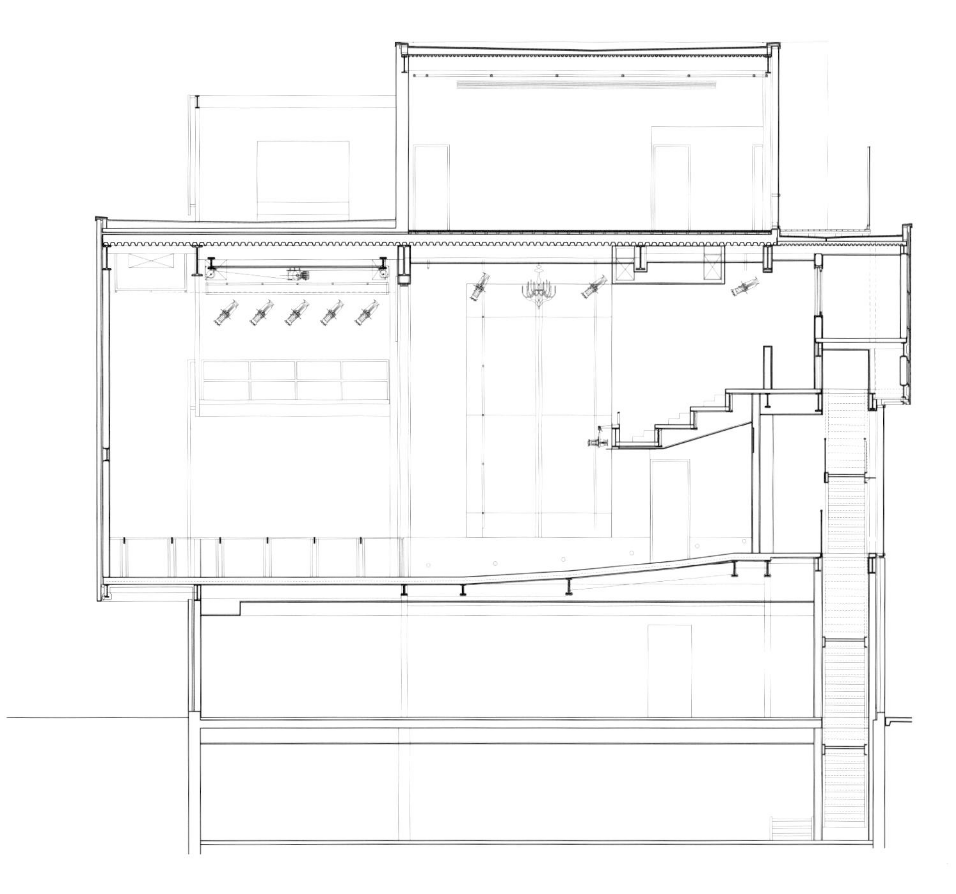

SORTIE

la cloche de
verre

条形建筑

办公建筑

Costa Mar Office Building

科斯塔玛办公楼

Design Company / 设计事务所:
RICARDO BOFILL TALLER DE ARQUITECTURA
Location / 地点:
Peru （秘鲁）
Size / 面积:
25 000 m^2

视觉亮点

设计师在建筑的外立面上设置了垂直的夸张的裂口，不仅巧妙地将空间一分为二，而且赋予了建筑奇特的外观。

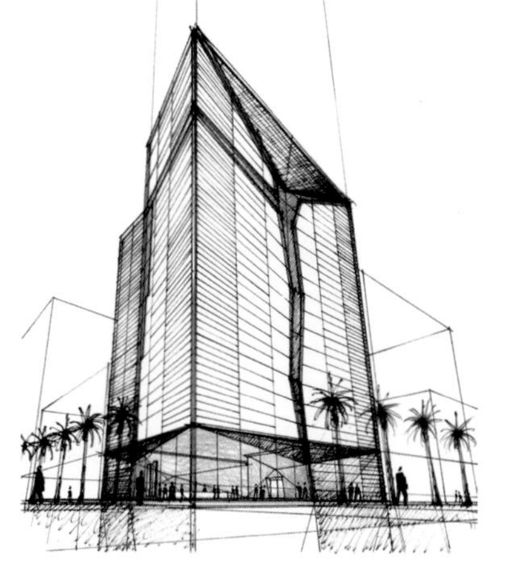

SAMSUNG
Tecnocom

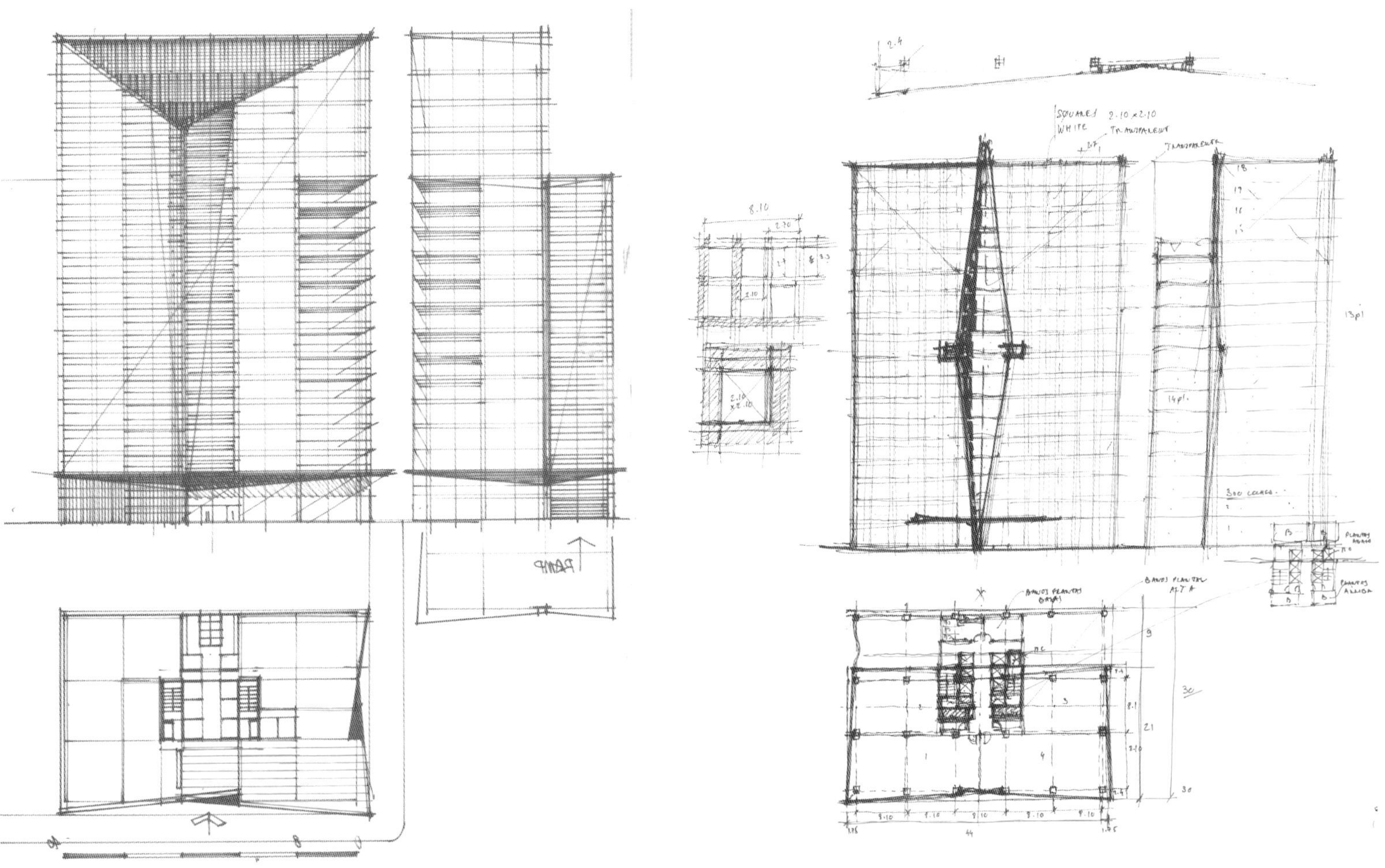

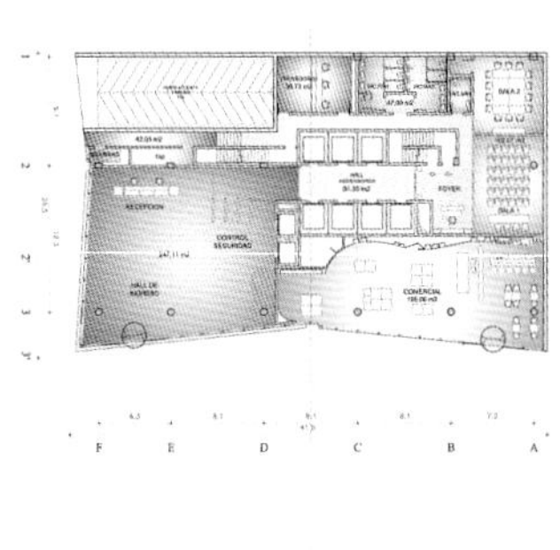

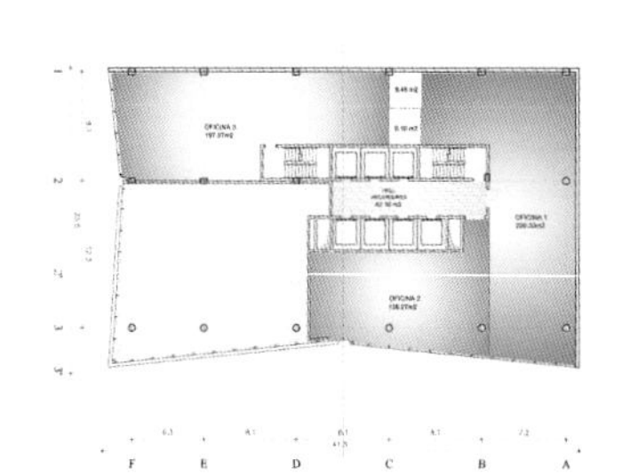

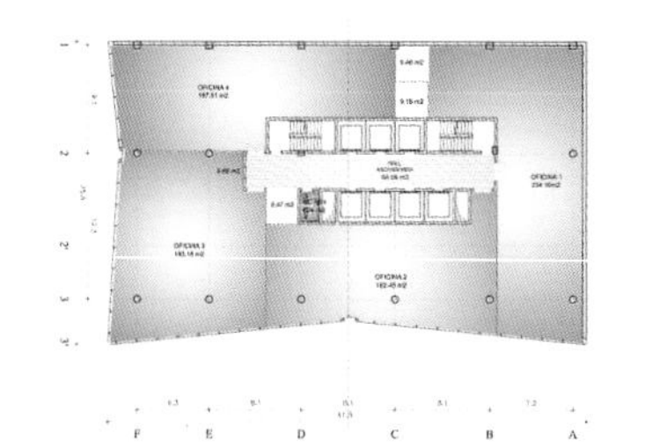

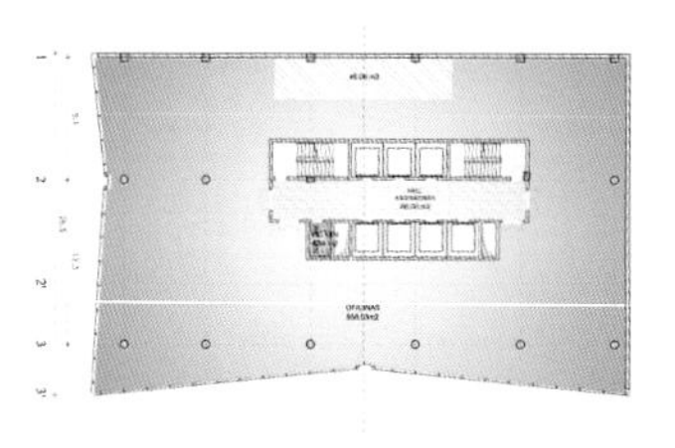

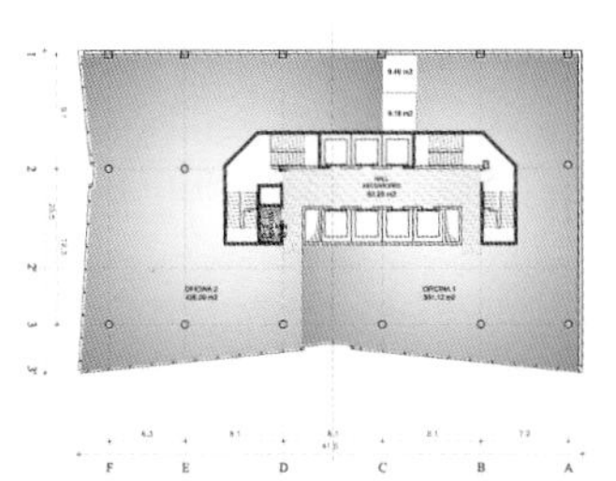

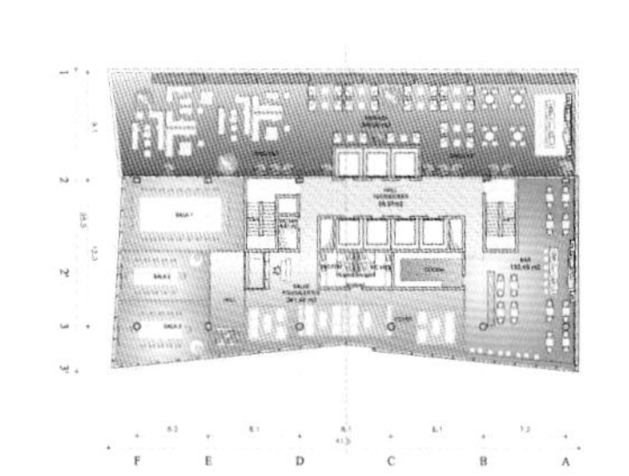

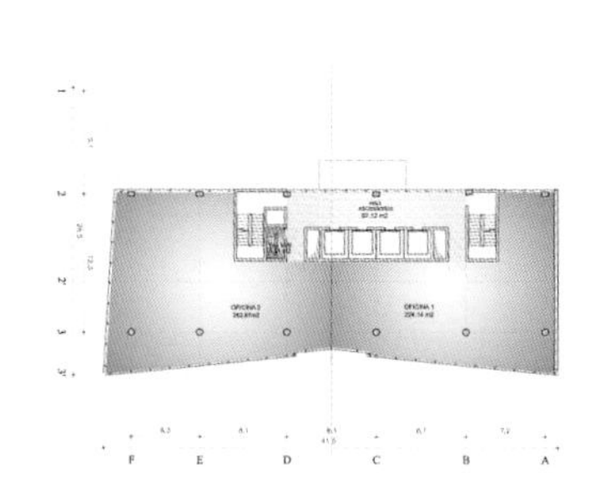

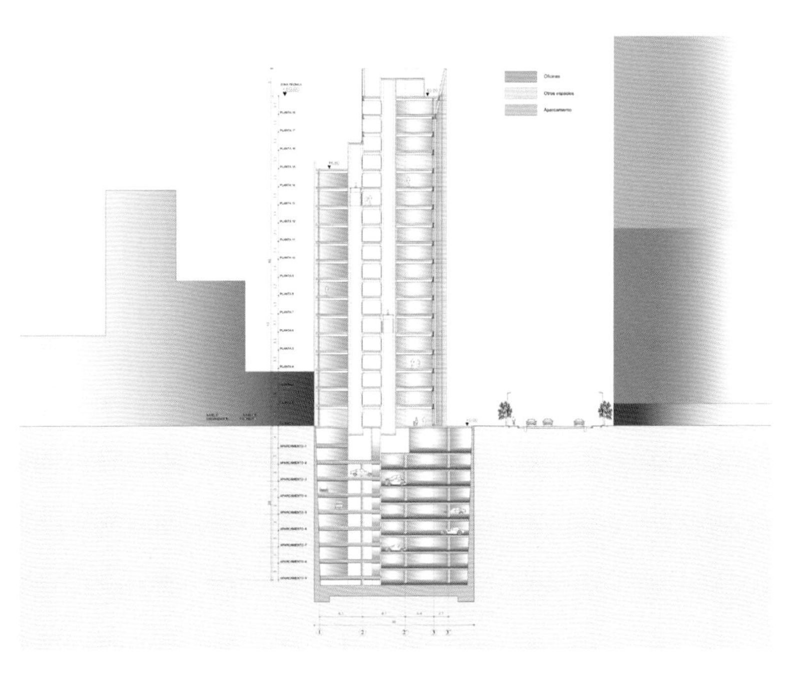

RBTA design for this iconic office building in San Isidro, Lima's business district expresses and encourages the dynamic activity of the area. The gate-shape of this corner building is emphasized by a vertical "scar" on the glazed main façade.

The scheme provides 15,000 sqm of flexible efficient office space over 18 floors and 10,000 sqm underground car park to accommodate 301 cars over 9 floors. The double height entrance lobby includes a small business centre, a shop and direct access to the car park.

Floors from 2 to 13 provide flexible space free of internal columns that allows subdivision into four offices of approximately 250 sqm. The load transfer at floor 14, allows the distribution into two large units. We have reserved a special distribution for floor 15 to accommodate a cafeteria with a large lounge terrace offering panoramic views of the city of Lima, board meeting rooms and reception area. Alternatively board meetings could be replaced by a fitness club. Top floors can accommodate one of two offices.

The serigraphed glass façade permits maximum natural daylight penetration thus reducing energy consumption.

RBTA 设计的这座办公建筑位于伊希德罗的利马商务区，是这个地区的标志性建筑。建筑立面上设置了垂直的像疤痕一样的裂口。

根据规划要求，建筑要提供超过 18 层的 15000 m^2 的有效的办公空间和 10000 m^2 的能容纳 301 辆汽车的超过 9 层的地下停车场。入口处需要设置小型商务中心、商店和进入停车场的入口。

从二层到十三层，要提供四处面积各约为 250m^2 的办公空间。十四层要分隔成两大空间。十五层要设置自助餐厅、董事会会议室（或者是健身俱乐部）、接待区，其中自助餐厅拥有欣赏城市风景的绝佳视野。建筑顶层必须容纳两个办公空间。

建筑的镜面材料为建筑提供了良好的采光条件，从而降低了室内的照明能源消耗。

条形建筑

商业建筑

Park Hyatt Hyderabad

海德拉巴柏悦酒店

Design Company / 设计事务所:
John Portman & Associates
Project Architect / 设计师:
John Portman & Associates, P.G. Patki Architects & T.V. Virani & Co. (local architect)
Location / 地点:
India （印度）
Area / 面积:
site 12 141 m^2 , gross building 55 700 m^2
Photography / 摄影:
Nicolas Dumont

视觉亮点

建筑外观井然有序，内部空间充满了自然气息。

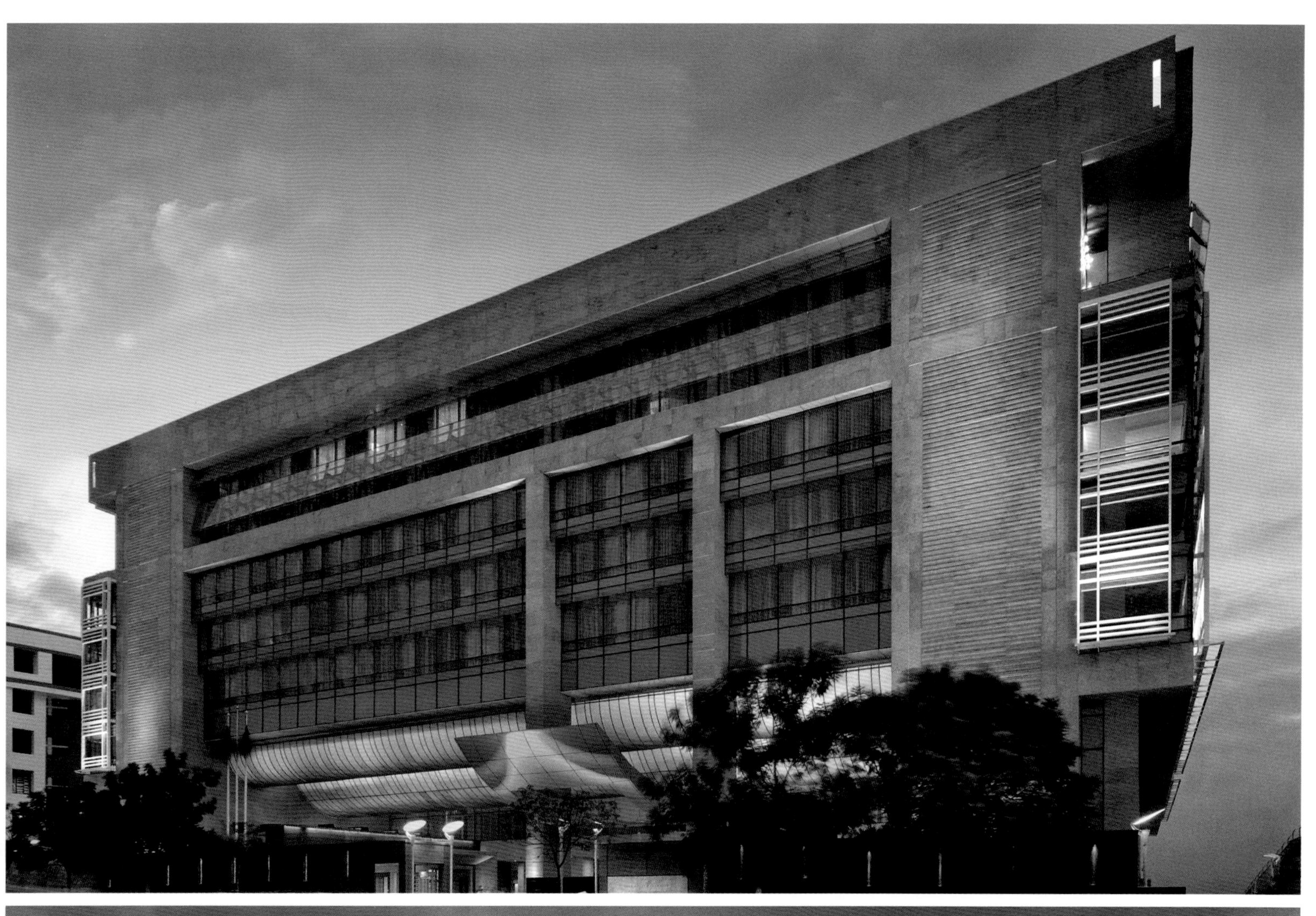

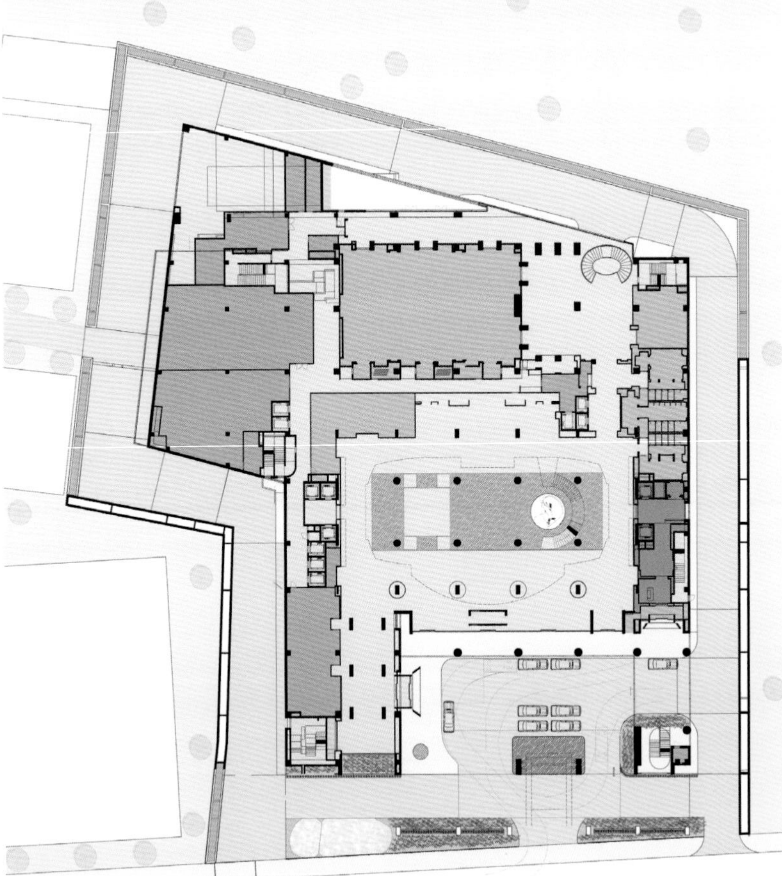

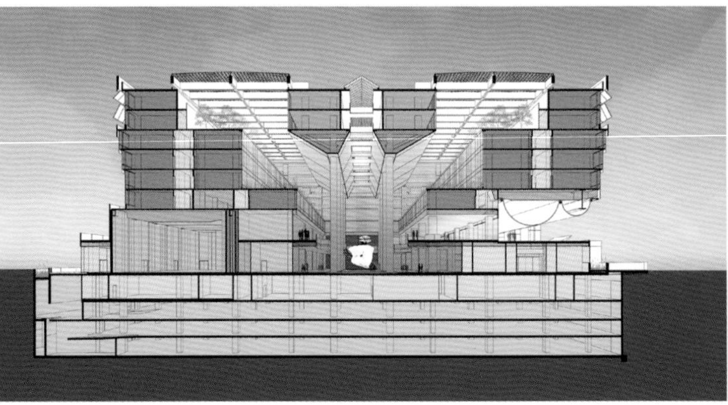

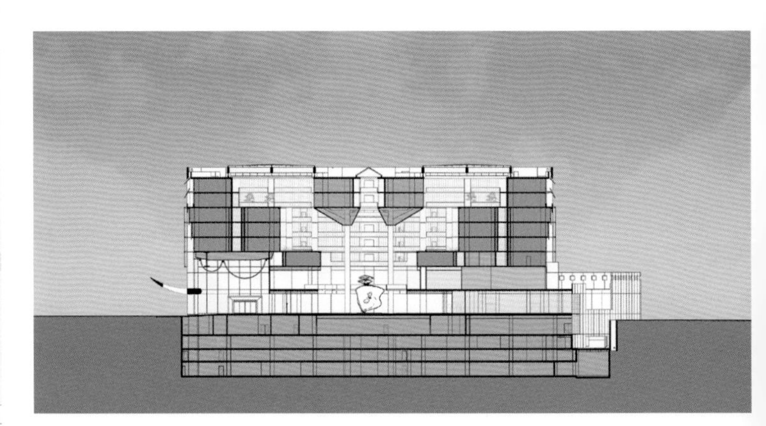

Hyderabad, the capital of Andhra Pradesh in south central India, has always been a bustling city and, with its recent emergence as a hub for the information technology industry, the rapidly growing metropolis has begun to expand across the area's arid terrain. The hot environment, coupled with the fast pace of India's fourth most populous city, led us to design the Park Hyatt Hyderabad as a place of cool, calm serenity and tranquility – a refuge from the intense traffic, harsh climate, and parched landscape.

Drawing inspiration from traditional Indian architecture, the design was conceived as a microcosm focused around an interior courtyard with terraced gardens, descending into a reflecting pool highlighted by a monumental sculpture that soars two stories high. Abundant interior plant life, the large reflecting pool, the openness of the atrium space, natural light, the color palette, the spa spaces, and air conditioning all contribute to the creation of an indoor oasis.

The sleek and modern eight-story structure is clad in natural Madurai granite from South India and features expansive glass windows. In addition to the indigenous granite, the façade elements acknowledge India's intense sun with the use of metal sunscreens, low E insulated glass and architectural accents.

南印度中部安得拉邦首府海德拉巴已经成为一座繁华的都市。这里气候炎热，生活节奏较快，而柏悦酒店为人们提供了一个凉爽、静谧的环境，使人们逃离那喧嚣的交通和酷热的气候。

该酒店设计从传统印度建筑中汲取了灵感，建筑内部空间就是一处微缩世界：空间布局以设有台阶的中庭为中心，中庭中设置了一处水池，水池中的最大亮点是那座两层楼高的雕塑。室内繁茂的植物、开放式的中庭、自然光、水疗馆、空调系统等创造了一个舒适的环境。

该建筑设计所面临的最大挑战就是以可持续、可操作的方式尽可能消减炎热和干旱的气候所带来的不良影响。井然有序而又时尚的八层建筑外部覆以原产于南印度马杜的花岗石，并设置了大型的玻璃窗。建筑立面通过设置金属遮阳板、绝缘玻璃等设施来阻挡印度南部强烈的光照。

条形建筑

商业建筑

Waldorf Astoria Shanghai on the Bund

上海外滩华尔道夫酒店

Design Company / 设计事务所:
John Portman & Associates
Project Team / 设计团队:
John Portman & Associates, Shanghai Architectural Design & Research Institute (SIADR) (local architect)
Location / 地点:
China （中国）
Area / 面积:
60 663 m^2
Photography / 摄影:
Michael Portman, Shu He Photo, Courtesy of Waldorf Astoria, Shanghai on the Bund

视觉亮点

建筑立面简洁、整齐，与街道上的原有建筑完美地融合在一起。

WALDORF ASTORIA
上海外滩华尔道夫酒店

AURORA

Like all Portman projects, the new Waldorf Astoria Shanghai on the Bund is designed to deliver an experiential, people-oriented place for the users and a solid return on investment for the owners and operators.

It was important to maintain the integrity of the waterfront of the historic Bund district. The program called for a guestroom tower, but great care was taken in the design and positioning of the tower so that its height does not dominate the historic club. Standing at street level, the focus is on the historic façade while the tower, by design, blends into the backdrop of the city.

The design of the spacious courtyard linking the new guestroom tower and historic club building was motivated by the desire to create an open public space for everyone to enjoy – a unique and engaging "people space" in the midst of the city.

In addition to being a welcoming public space, the courtyard is multifunctional to accommodate both hotel use and public cultural events. Landscaping and water features complement the various functions and enhance the sensory experience of the open space environment.

就像波特曼设计事务所承建的其他项目一样，位于上海外滩的华尔道夫酒店致力于为入住酒店的客人们营造一处人性化的空间，并为酒店业主和经营者带来切切实实的经济回报。

对于该项目而言，最重要的一点是设计要确保历史悠久的外滩滨水区的完整性。设计师在建筑的空间设计和建筑定位方面也花费了很多心思。当人们处于建筑一侧的大街上时，焦点集中在这座建筑的立面上，而酒店建筑本身与整座城市的背景融为一体。

设计师设计的开阔的中庭将新建的客房大厦与上海历史悠久的建筑联系起来，营造了一处令人倍感惬意的开放式公共空间——位于城市中心的具有独特魅力的“人性化空间”。

除了作为一处吸引人前往的公共空间外，中庭还是一处多功能的空间，既有酒店功能，又可举办一些公共文化活动。水景的设置使空间功能更显多样化，同时提升了人们对开放式空间的感官体验。

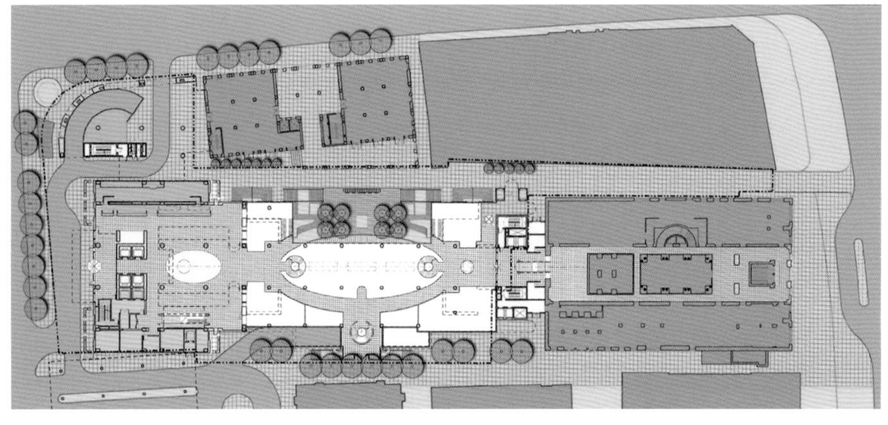

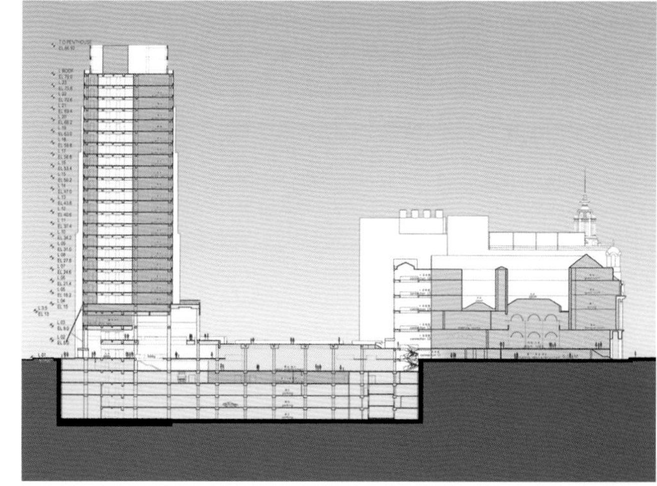

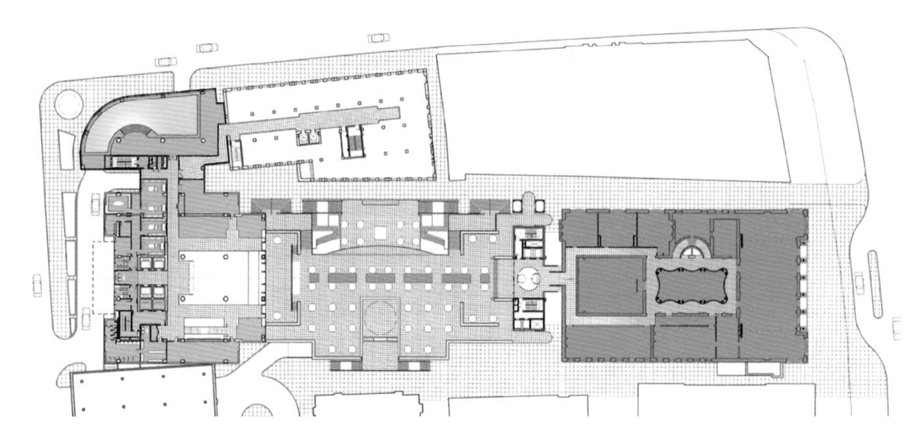

条形建筑

商业建筑

Chungha Building

首尔商业大楼

Design Company / 设计事务所:
MVRDV
Location / 地点:
South Korea （韩国）
Area / 面积:
2820 m^2
Photography / 摄影:
Kyungsub Shin

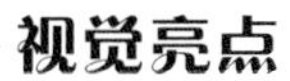

视觉亮点

白色的、亮眼的大幅广告贴面，使整座建筑重新活跃起来。

PRADA
LOUIS QUATORZE

PRADA
PRADA

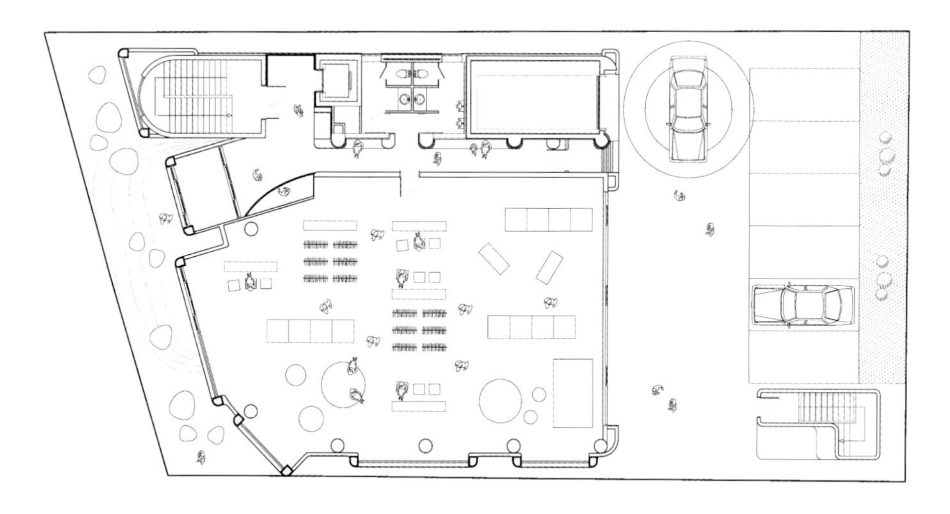

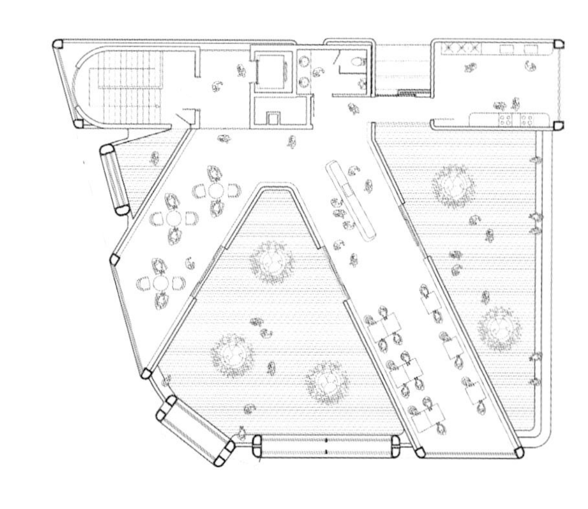

Just before a Korean pop-song became a global success on YouTube for the first time in history, MVRDV was commissioned by Woon Nam Management Ltd. to redfine a building on Gangnam's chic Apgujung Road. Even though the Chungha building was completed in the 1980's it was already outdated in a street dominated by flagship stores.

The Chungha building had become a rotten tooth in a fast changing streetscape dominated by single brand stores, this building contains a collection of brands in one. On the previous façade, a motley collection of fonts competed for the attention of passersby. The sober building's beige natural stone façade was ruined by commercial messages. The ground floor is occupied by French leather accessories label Louis Quatorze, the floors above hold a wedding planners' office, the clients' maintenance society and two plastic surgery practices. The windows of these floors were hermetically sealed, adding to the worn out feel of the structure.

The new façade concept is convincingly simple: Chungha is a multiple identity building which was transformed into a collection of shop windows so each commercial venture imposed onto the façade would have a fitting canvas for its display. Curvaceous frames were found to be the best match to the large amount of shop windows, and a mosaic tile consequently became the façade material to follow the curves. LED lights change the buildings appearance. MVRDV was given nine months to complete the refurbishment. Adding to the complexity was the limited size of the construction site - five storeys tall but only 2,5 metres at its widest point.

自从《江南 style》这首流行音乐在全球一举成名后，韩国江南区便成为人们关注的对象，其中 Woon Nam 公司委托 MVRDV 设计事务所把建于上世纪 80 年代的旧总部大楼改造一番。仅仅 9 个月，这栋建筑便焕然一新地出现在已经拥有众多旗舰店的大街上。

原有的这座建筑和它周边的时髦旗舰店相比，有一些难看。这里面进驻了许多品牌：一楼是法国皮革配饰品牌 louis quatorze，上面几层是婚礼筹划公司的办公室、业主权益维护办公室及两个整形外科诊所。建筑浅褐色的自然石材立面被租户们的商业广告、招牌和信息弄得面目全非。

新的外立面设计较为简单，大小不一的整面窗户可以让商户尽情展示自己的商品和海报。新立面有着白色的、亮眼的马赛克贴面，这些贴面近看像泡泡，远看如同整块光滑的石材，而且能呈现完美的曲线造型。窗户内部安装有 LED 灯，在夜间发出变幻诱人的光芒。该设计事务所 MVRDV 用 9 个月完成了建筑的改造。建筑共有 5 层，但最宽的施工宽度仅 2.5m。

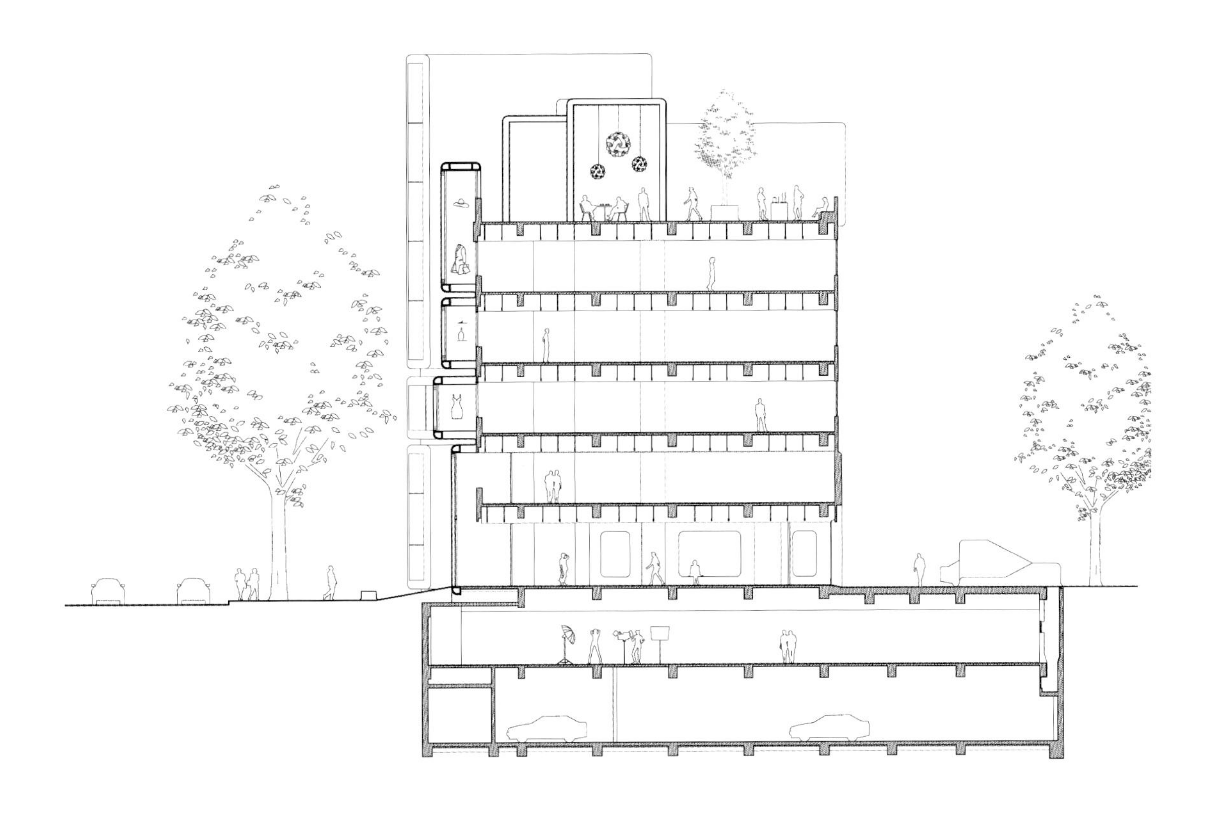

SAVILE ROW
LOUIS QUATORZE
PRADA

아트성형외과
골든뷰
성형외과
골든뷰 성형외과
S AVILE ROW
PRADA
LOUIS QUATORZE
LOUIS QUATORZE

비포앤애프터
GYM
하지운
성형외과
NK Clinic
LOUIS QUATORZE

골든뷰성형외과
LOUIS QUATORZE

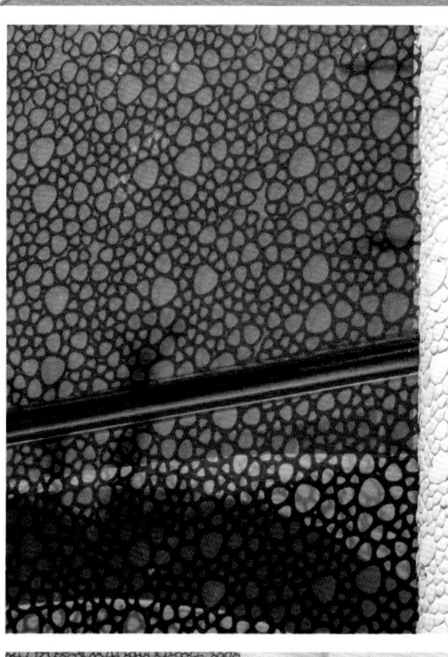

条形建筑

办公建筑

B & Q Store Support Office

百安居集团办公楼

Design Company / 设计事务所:
BDP
Location / 地点:
UK（英国）
Photography / 摄影:
David Barbour, Sanna Fisher - Payne

视觉亮点

建筑立面由多个长方形构成，建筑外观简洁、明晰，大面积的玻璃材料为室内提供了明亮而通透的办公环境。

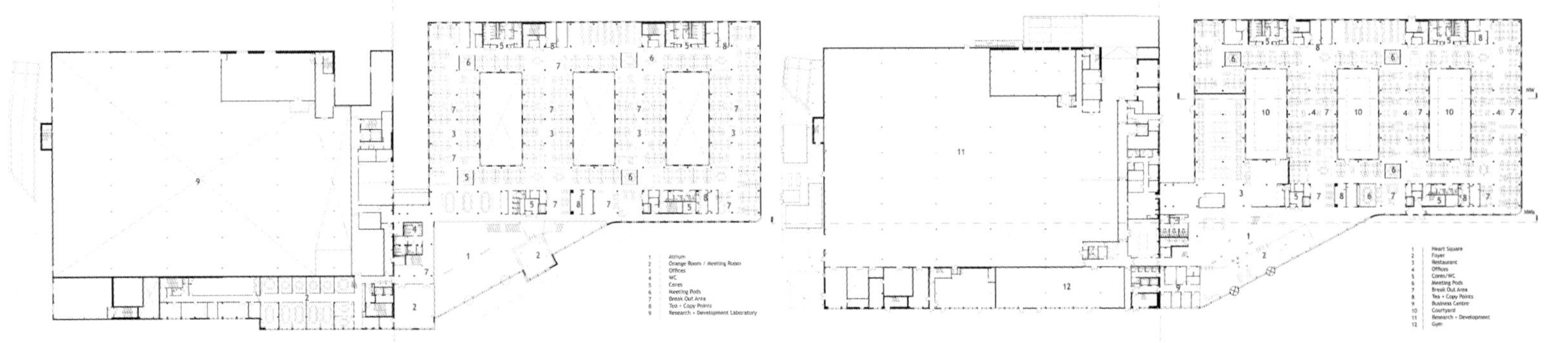

B&Q's new store support office building in Eastleigh, Hampshire, was designed by an interdisciplinary BDP team following its principle of 'creating places for people'. The building comprises two main parts, the research and development laboratory and the offices. Built in two phases, the office spaces correspond with the previously established organisation of spaces introduced by the laboratory. The building envelope is also an architectural response to this spatial organisation – the north-east and south-east facades are more public facing, whereas the north-west and south-west facades are primarily for services use.

The main office space works on a 13.5m structural grid floorplate organised around three internal courtyards. The offices have been designed around the principles of maximising use of natural daylight and ventilation, whilst providing a highly efficient operational environment for B&Q. The use of the courtyards allows for interaction between the office space and external landscape, and at the same time maximise infiltration of natural light to the office floor plate.

The office areas of the building have a glass curtain wall with solid parts to deliver optimum transparency and solar protection. During the day the facade allows controlled natural daylight to enter the building and at dusk the internal lighting means that the building 'glows' in its surroundings.

百安居集团在英国汉普郡的伊斯特雷格区新建了一座办公大楼。该建筑由两部分组成，即研发实验室和办公区。整个项目建造过程分为两个工期，办公区的设计遵循之前确定的空间架构。建筑的外部设计也与内部空间相呼应。

主要办公区围绕着三处中庭空间展开设计。办公空间的主要设计原则是创造良好的采光和通风条件，同时为百安居集团提供一个高效的办公环境。大面积中庭的设计实现了外部景观环境在办公空间的延续，同时使尽可能多的自然光进入办公空间。

建筑办公区域的外墙上用玻璃和实木板做装饰，这样的设计使内部空间具有良好的通透性，同时避免内部空间遭受日光暴晒。白天，自然光穿过外立面进入建筑内部；夜晚，建筑内的照明系统使整座建筑成为“发光体”。

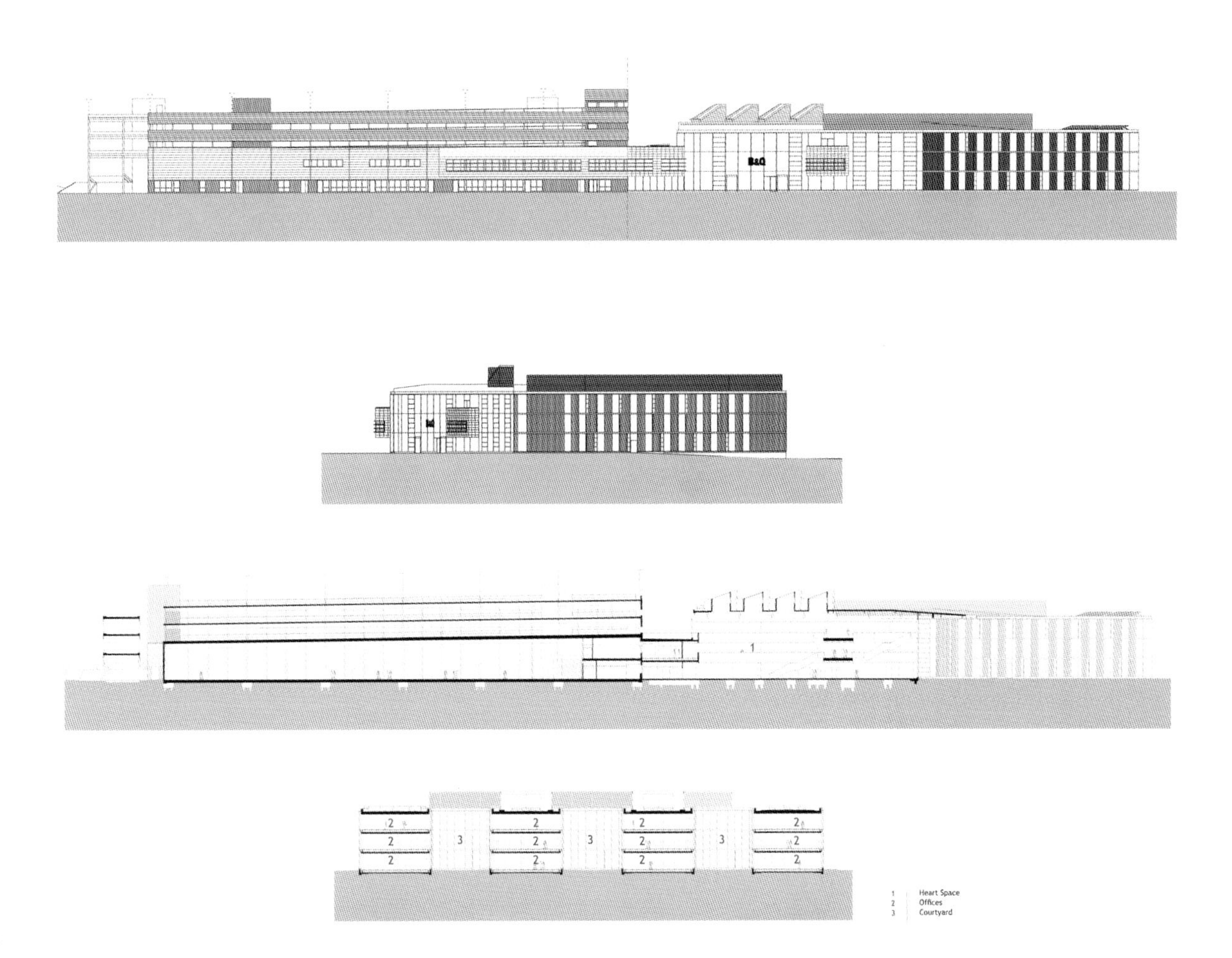
B&Q
1 Heart Space
2 Offices
3 Courtyard

Siège EDF - Ajaccio

白色盒子

Design Company / 设计事务所:
ECDM
Location / 地点:
France（法国）
Area / 面积:
3752m^2

视觉亮点

建筑就像是包裹在泡沫塑料里的盒子。建筑外观简洁、优雅，具有强烈的秩序感。

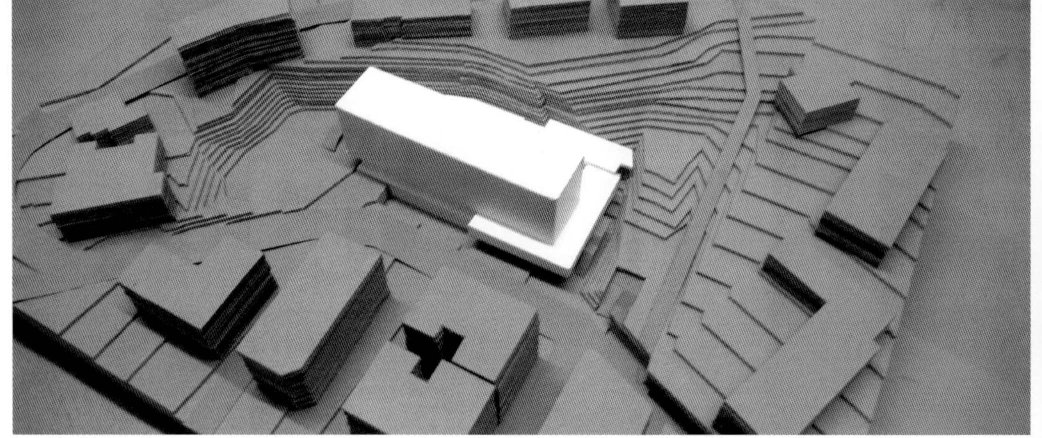

C.H.G
MISERICORDE
Centre des Impôts
Clinique Sud

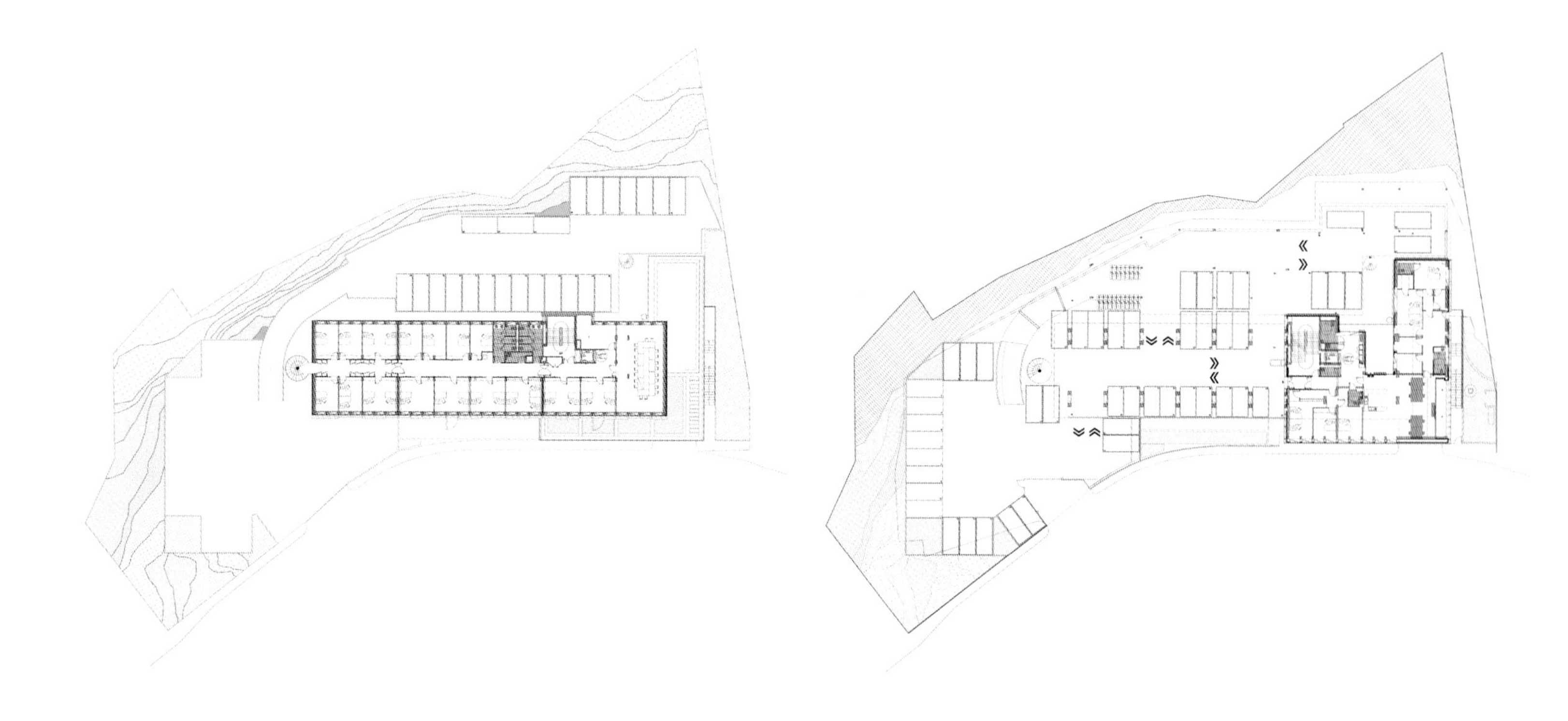

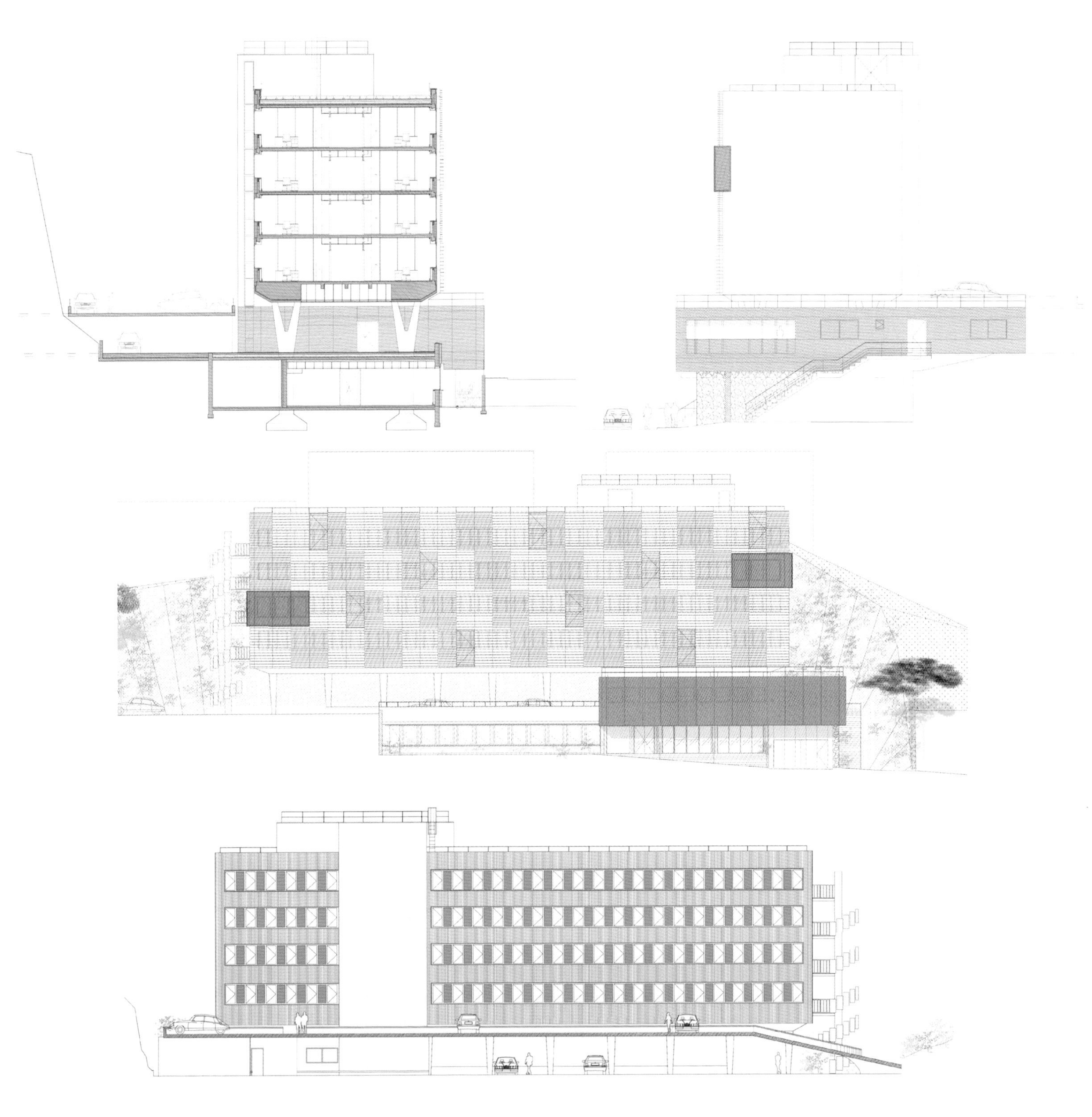

Built in 1969, the EDF has a great architecture, representative of modernity 60s. It consists of three volumes, and has, because of its location on the slope, two levels of ground floor, said lower floor and upper floor.

Our intention is to strengthen reading the simple and elegant massing. It is a building of a certain presence and we believe it is important to highlight the beautifully crafted the existing architecture by placing us in the conceptual original party . We propose a unified treatment of the facades, while reinterpreting the concept of sunscreen already in place. South facade, overlooking the avenue Eugenie, is equipped with horizontal strips of lacquered aluminum, fully covering the main facade facing south. This principle allows giving a unit volume of the building. By the openings in the background, the front scales, and offers a softer perception of volume, in keeping with its surroundings and the local landscape. Subtle shifts in the frame blades create vibrations of the facade and mitigate the effect of a mass reading relief, contrasts and flicker.

原有的建筑建于 1969 年，是一座雄伟的建筑，体现了 20 世纪 60 年代的建筑风格。整座建筑由三部分组成。基于该建筑所处的是斜坡地块，建筑主体被分为下层结构和上层结构。

设计师的主要设计意图是强化其简洁、优雅的建筑外观。设计师认为比较重要的一点是通过对原有的建筑格局进行充分研究，进一步突显原有建筑精彩的设计之处。设计师提出要将建筑立面作为一个整体来看待，同时对原先就有的遮光屏进行重新设计。南部立面正对着一条大街，这个立面上都设置了水平漆制的铝板材。建筑与周边环境、当地景观完美相融。框架板材巧妙的移位处理也使整个立面具有无限的活力，使建筑外观更显生动。

条形建筑

National Theatre 'The Shed'

英国国家大剧院临时场馆

Design Company / 设计事务所:
Haworth Tompkins
Location / 地点:
UK（英国）
Area / 面积:
628 m²
Photography / 摄影:
Helene Binet, Philip Vile

视觉亮点

建筑就像是一个放倒了的方凳。红色的建筑外墙十分显眼。无开口的外墙设计赋予建筑神秘的色彩。

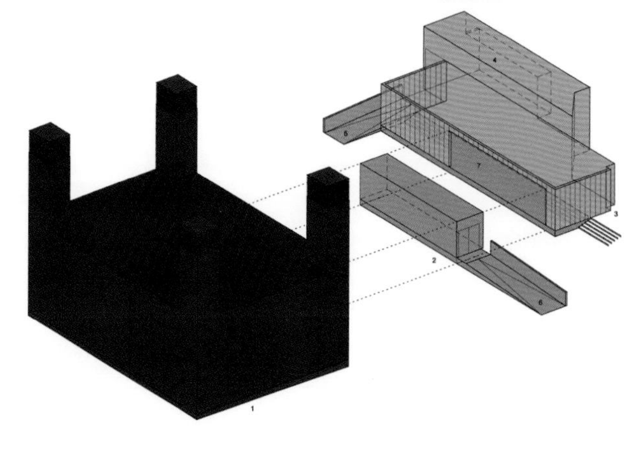

National

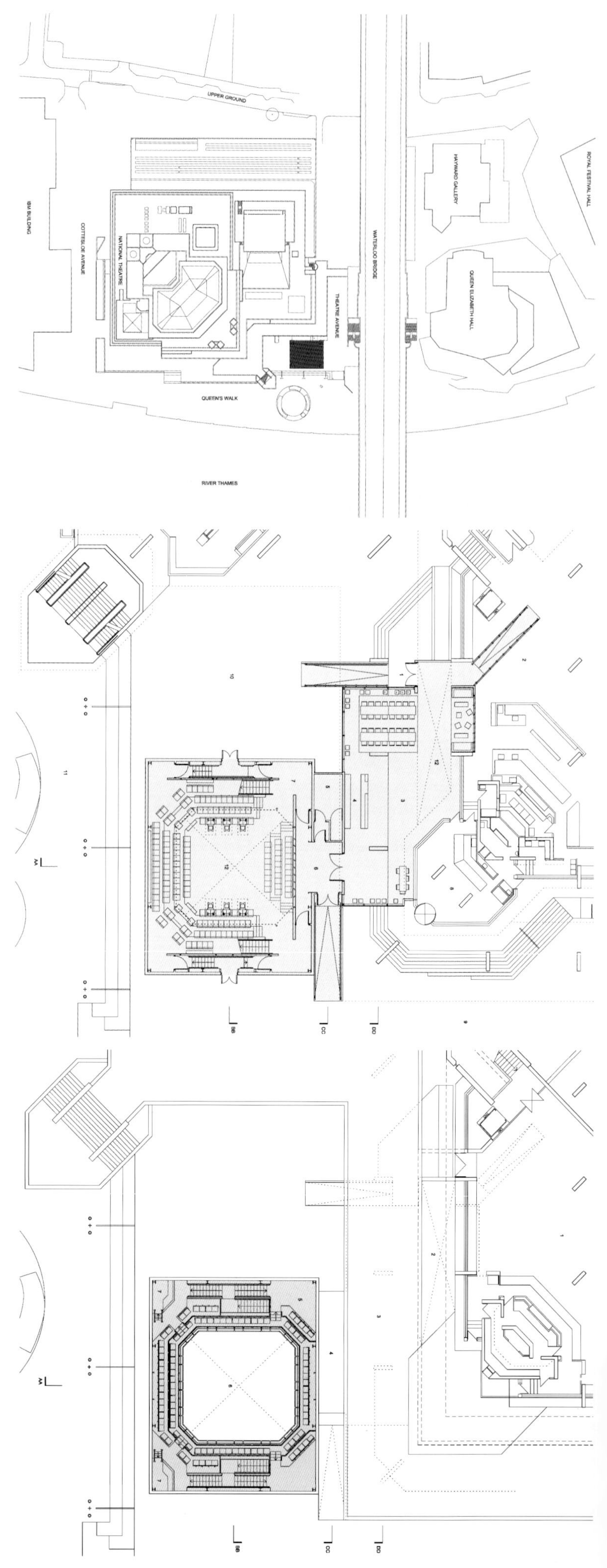

The Shed occupies Theatre Square, at the front of the National Theatre, beside the river. Its simple form houses a 225-seat auditorium made of raw steel and plywood, while the rough sawn timber cladding refers to the National Theatre's iconic board-marked concrete, and the modelling of the auditorium and its corner towers complement the bold geometries of the NT itself. A temporary foyer has been carved out from the space beneath the NT's external terraces and provides easy connection to the existing foyers. The Shed's brilliant red colour covering the entire mass of a form without doors or windows, announces its arrival boldly against the concrete bulk of the NT, giving it a startling and enigmatic presence.
The Shed also represents another step in Haworth Tompkins' ongoing project to research sustainable ways of making theatres. Built of materials that can be 100% recycled and fitted out with re-used seating, The Shed is naturally ventilated, with the four towers that draw air through the building providing its distinctive form.

这座临时场馆紧邻英国国家大剧院，位于泰晤士河南岸。该建筑由粗钢和胶合板打造而成，其中设置了一座可容纳 225 人的礼堂。其粗犷的木质外墙与国家大剧院标志性的模纹外墙相呼应。设计师在大剧院外部露台的下方设置了一处临时门厅，并设置通道与临时场馆相连接。临时场馆的整个外表面都饰以耀眼的红色。设计师巧妙地将入口区设置在了大剧院的混凝土结构中，为临时场馆营造了神秘的氛围。
这座建筑秉承了可持续发展的设计理念，临时场馆的建造材料和剧院中所有的座椅 100% 可回收利用。建筑还可实现自然通风，建筑四角的结构不仅将自然风引入建筑内部空间，还赋予整座建筑非常别致的外观。

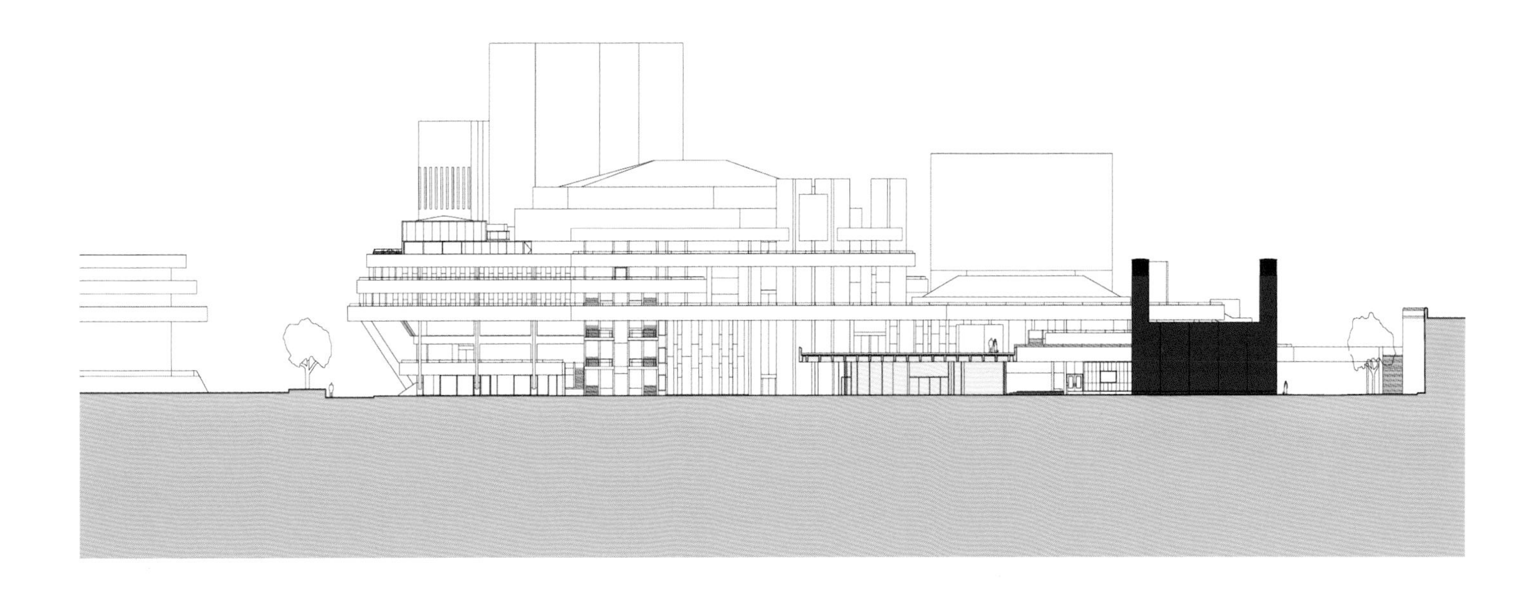

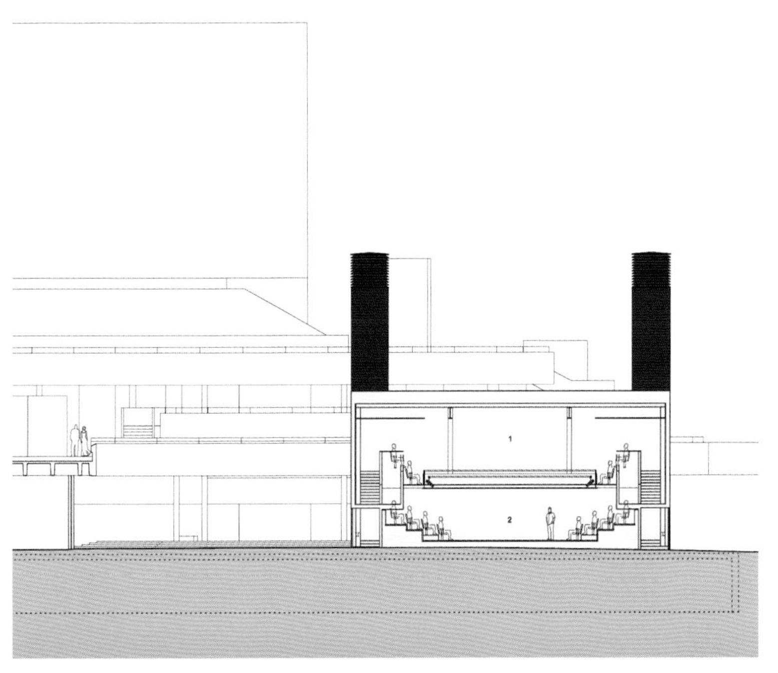
1
2

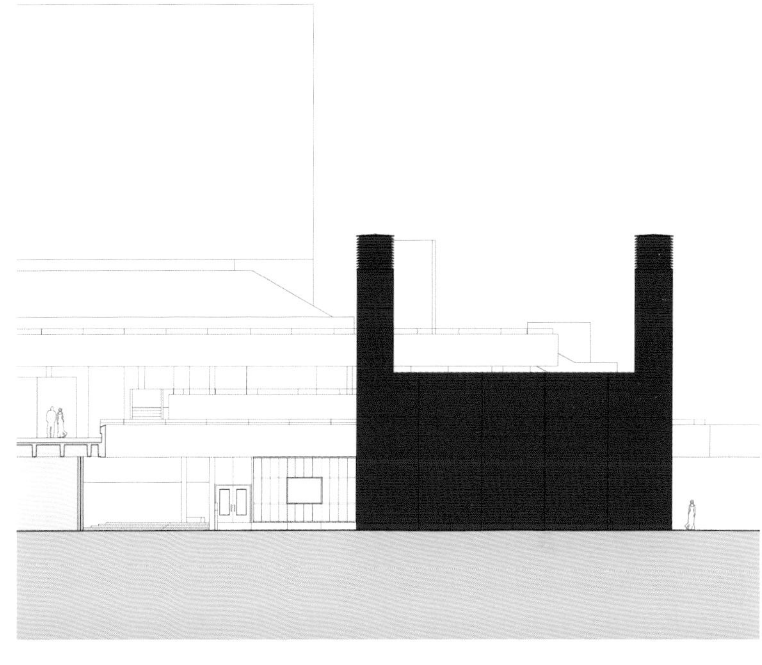

Theatre

条形建筑

文化建筑

Nebuta House

Nebuta文化中心

Design Company / 设计事务所:
molo with d/dt Arch, Frank la Rivière Architects Inc
Project Architect / 设计师:
Todd MacAllen, Stephanie Forsythe
Location / 地点:
Japan （日本）
Area / 面积:
site 13 012 m^2, building 4 340 m^2, gross floor 6 708 m^2
Photography / 摄影:
molo

视觉亮点

红色的条形钢带是建筑外立面的主要元素。

東横イン
東横イン

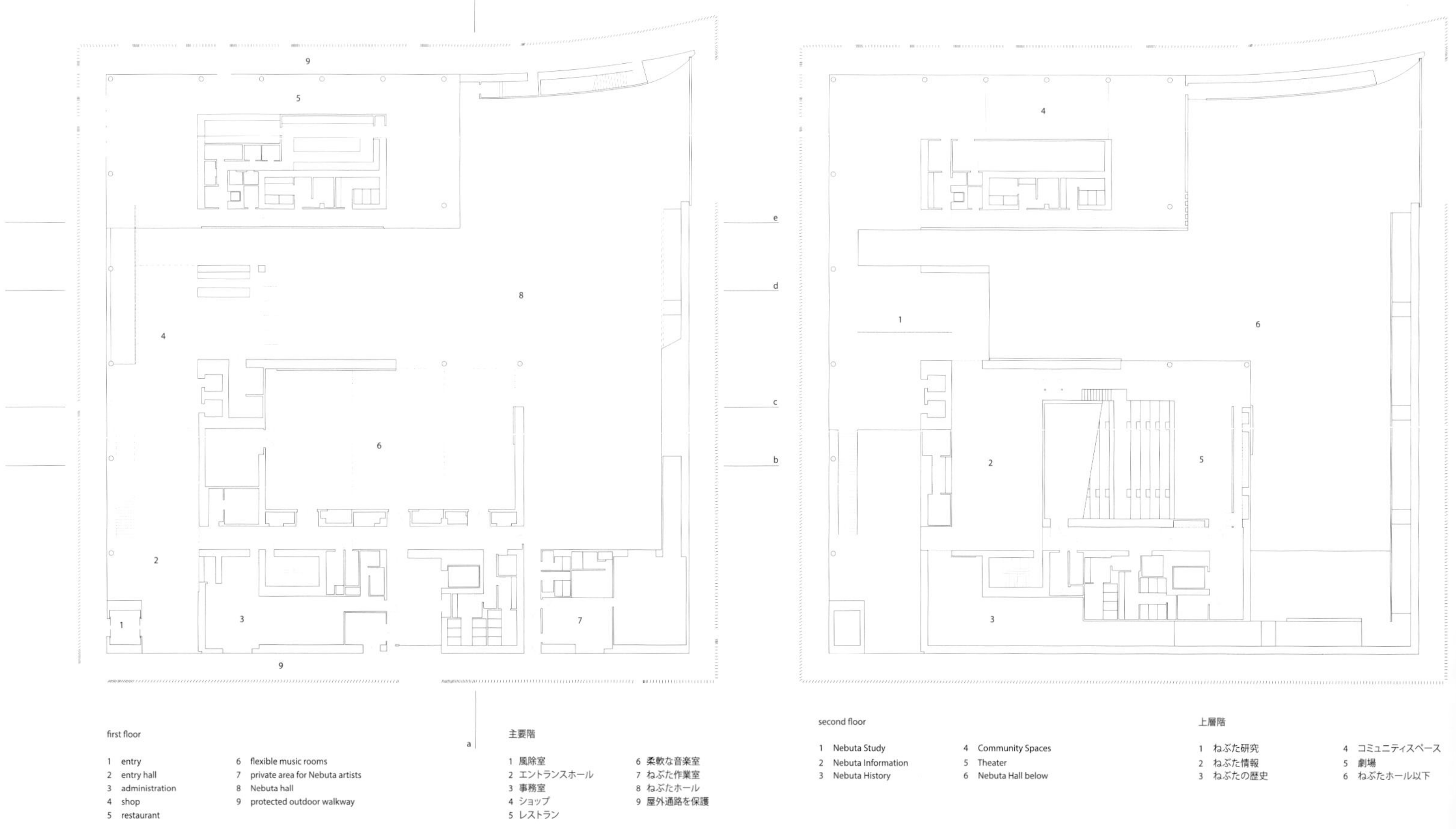

first floor

1 entry
2 entry hall
3 administration
4 shop
5 restaurant
6 flexible music rooms
7 private area for Nebuta artists
8 Nebuta hall
9 protected outdoor walkway

主要階

1 風除室
2 エントランスホール
3 事務室
4 ショップ
5 レストラン
6 柔軟な音楽室
7 ねぶた作業室
8 ねぶたホール
9 屋外通路を保護

second floor

1 Nebuta Study
2 Nebuta Information
3 Nebuta History
4 Community Spaces
5 Theater
6 Nebuta Hall below

上層階

1 ねぶた研究
2 ねぶた情報
3 ねぶたの歴史
4 コミュニティスペース
5 劇場
6 ねぶたホール以下

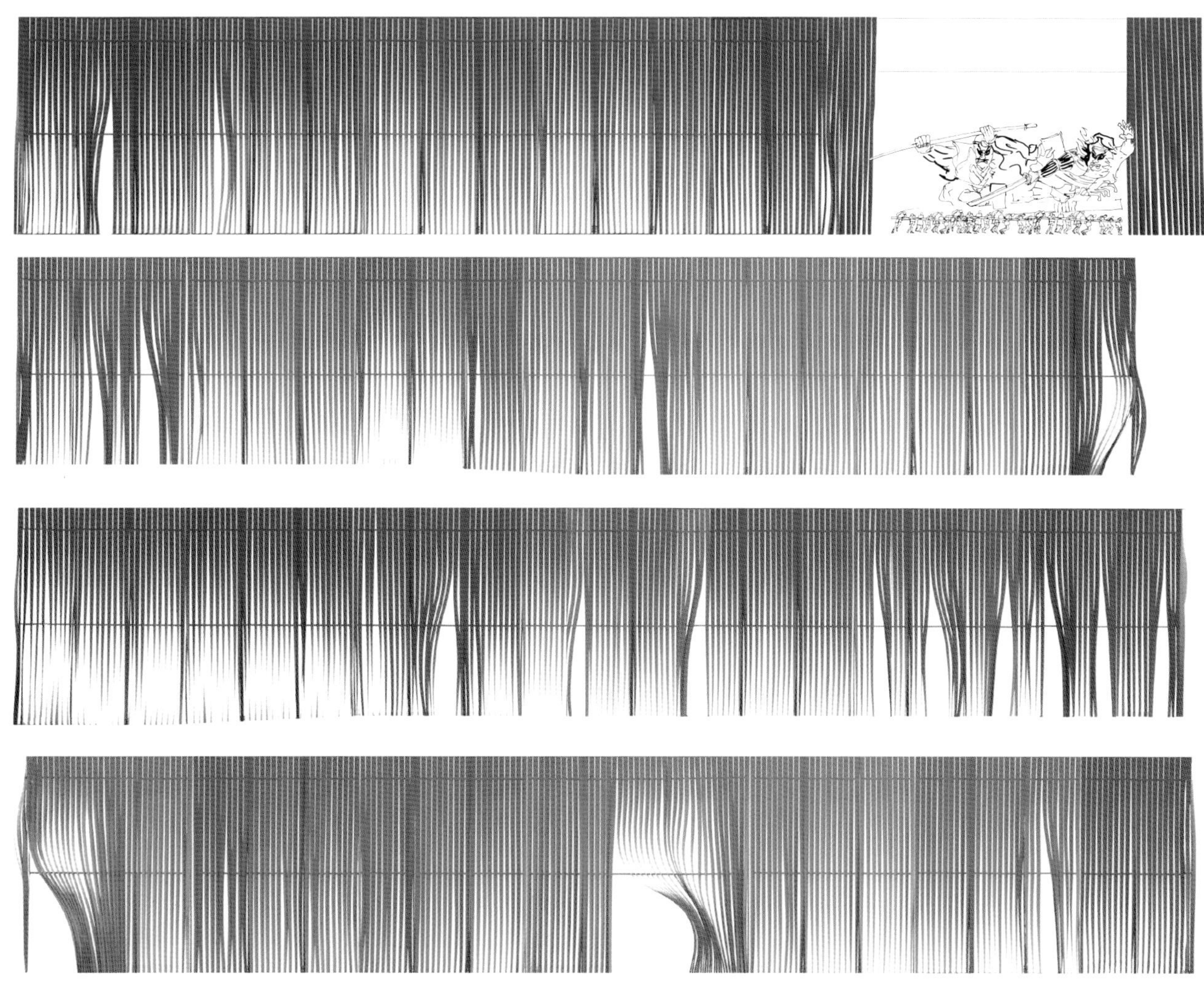

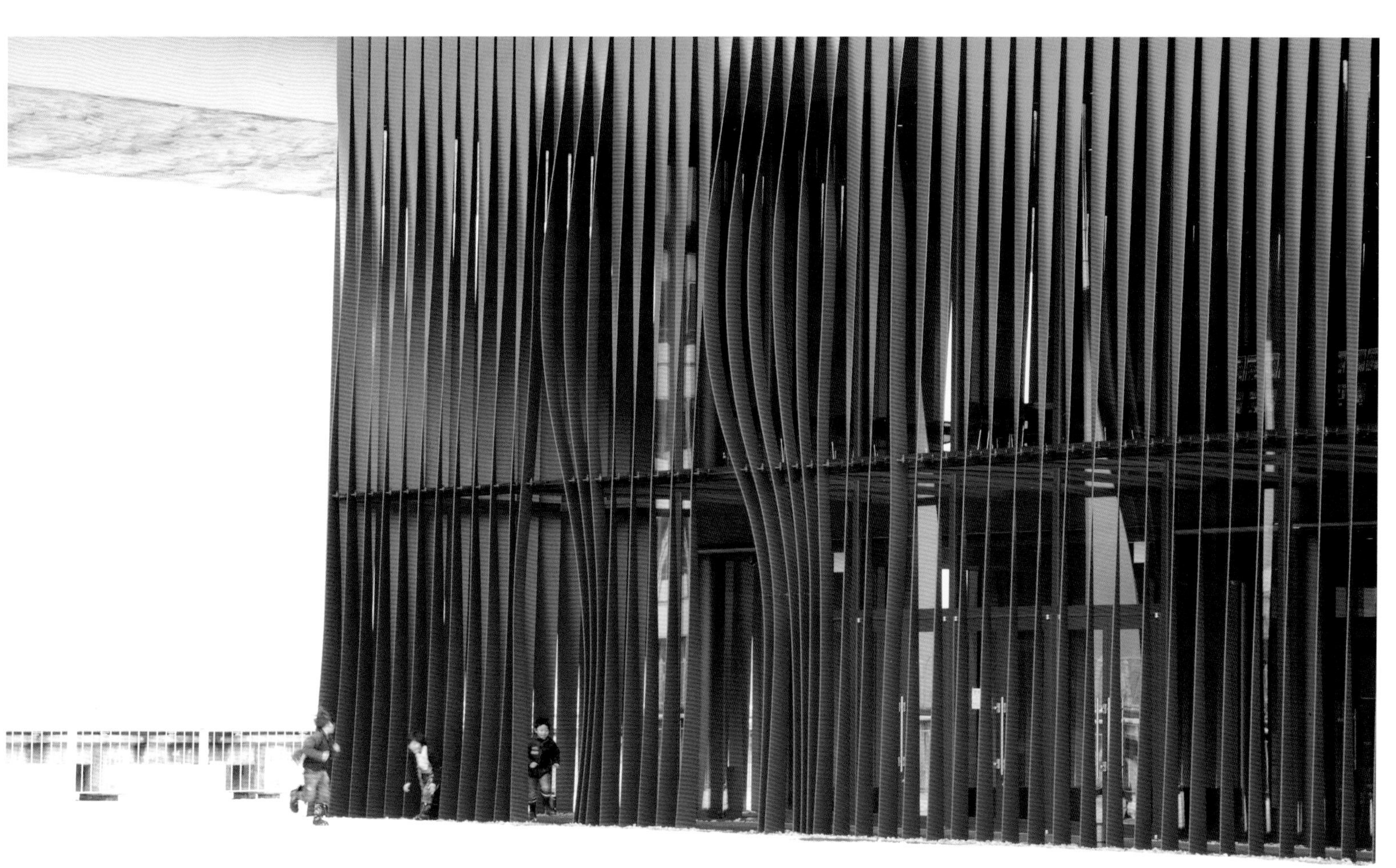

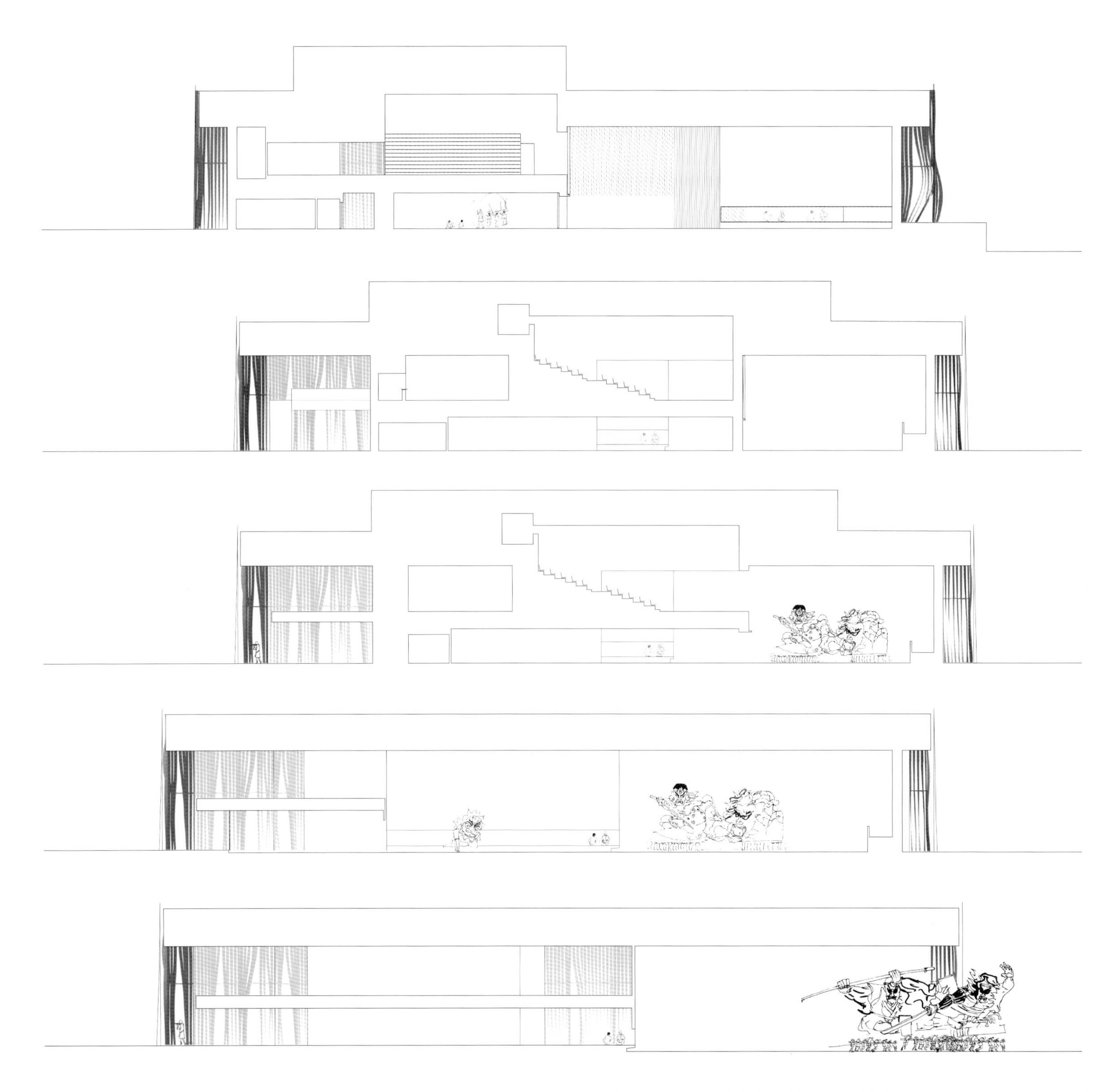

The building is enclosed by ribbons of twisted steel, enamel-coated deep vibrant red and individually shaped to create variation: openings for light, areas of opacity, views, or opportunities for pedestrian circulation. For each steel ribbon, the bottom was set to a unique and specific angle, with thought to how sunlight would permeate the ribbons as it moved throughout the day, while the top part of each ribbon remains parallel to the building.

The ribbon screen façade creates a sheltered outdoor perimeter space called the "engawa", a spatial concept originating in traditional Japanese houses. In this case, a dwelling for giant paper heroes, demons and creatures, the engawa acts as a threshold between the contemporary world of the city and the world of history and myth. Shadows cast on the walls and floor through the exterior ribbons have the effect of creating a new material. Shadow and light become another screen – the convergence of material, light, shadow and reflection changing with the sun and weather.

整座建筑外墙被深红色的钢带所包围。钢带产生扭曲进而形成开口，有的开口设置是为了实现自然通风并引入自然光，有的是为了人们更好地欣赏户外的风景，还有的被设计为入口。每一条钢带的底部都有特别的角度，以确保阳光能够进入室内空间，并在一天中发生美妙的变化。

钢带与建筑表面之间的有一个被遮蔽的户外空间，这个空间类似于日本传统民居的廊道。博物馆在这里充当神话中英雄、恶魔、生灵的居住地，而这条廊道则是他们所处的魔幻世界与现实世界的分界门槛。光线透过钢带反射在墙壁和地面上，形成美妙的光影效果。这种光影构成了新的景象——随光照、气候变化的光、影、映像的集合。

文化観光交流施設
ねぶたの家 ワ・ラッセ
青森市文化観光交流施設 ねぶた

Ayasha Building

安亚莎建筑

Design Company / 设计事务所:
METROPOLIS
Design Team / 设计团队:
Jose Orrego, Alicia Zapata
Location / 地点:
Colombia （哥伦比亚）
Area / 面积:
6 500 m^2
Photography / 摄影:
Claudia Paz

视觉亮点

多个LED灯带的加入使条形的建筑立面富有动感和活力。夜幕降临，建筑黑色玻璃结构就像消失了一般，呈现的只是那些灵动、时尚的线条。

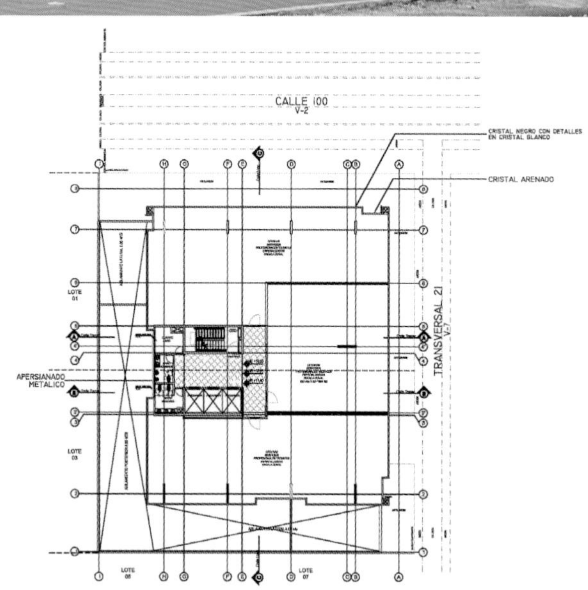

CALLE 100
V-2
CRISTAL NEGRO CON DETALLES EN CRISTAL BLANCO
CRISTAL ARENADO
APERSIANADO METALICO
TRANSVERSAL 21
V-7

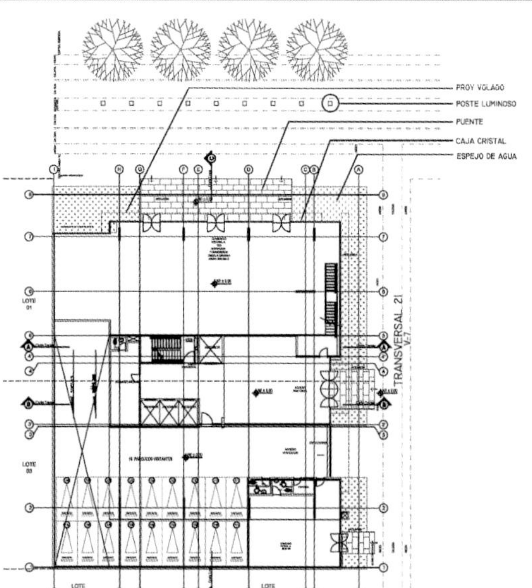

PROY VOLADO
POSTE LUMINOSO
PUENTE
CAJA CRISTAL
ESPEJO DE AGUA
TRANSVERSAL 21
V-7

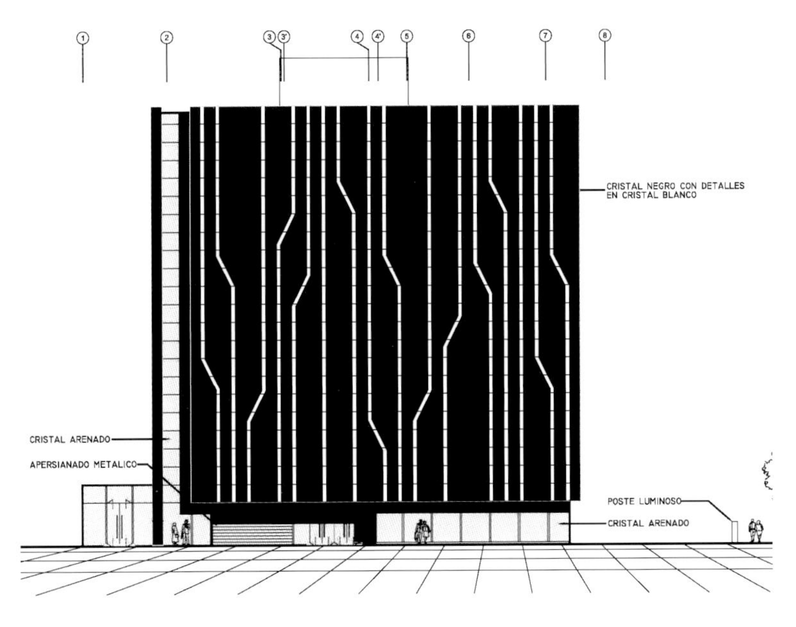

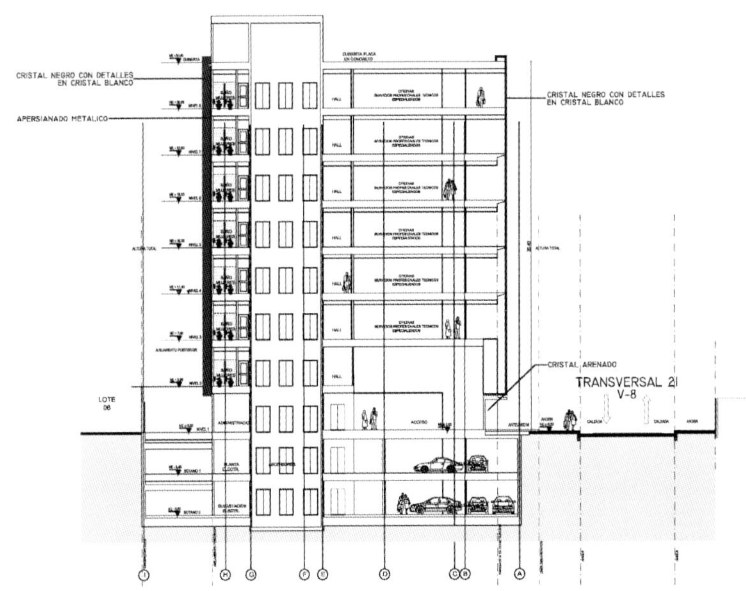

A new corporative building was the objective of the contest organized by a Colombian real estate developer, won by peruvian architect José Orrego. The project tried to emphasize its iconic character, standing out among the urban context. Differing from other brick and concrete Colombian buildings, this one aspired to be a four-surface well defined volume. Each plane is conformed by a fragmented black pane of glass. The LED lines that cross this big surfaces provide a tech and contemporary look. These lines are the result of the contrast of two types of glasses: black and frosted.

On the edges, planes are barely detached from each other by a luminous burnish that gives us the sensation of floating pieces. At night, the LED lines sick out with a programmed system, so that the whole building acquires a special dynamic and continuous movement. The black pane of glass dematerializes. The basis of this complex is a water mirror that reflects the building effects and creates a vertical symmetry. AYASHA means "small" in native language. In spite of that, the building shows itself as a huge urban referent.

该项目由哥伦比亚的一家建筑事务所担当设计。该项目设计的主要目标是设计一座具有标志性外观的建筑，使其区别于周边的城市建筑。不同于其他用砖石、混凝土打造的哥伦比亚传统风格的建筑，该建筑轮廓分明、风格独特。建筑的每个立面均由黑色玻璃板材打造而成。建筑的各个立面上都设置了条形 LED 灯带，这些灯带不仅具有照明作用，更赋予整座建筑时尚的外观。

夜幕降临，通过设定好的程序，LED 灯带开始运作，赋予整座建筑以特别的动态美感和连续式的移动观感。黑色玻璃板结构就像消失了一般。下方的水面倒映着整座建筑。“Ayasha”在当地语言中有“小”之意。尽管如此，该建筑成了一座城市地标性建筑。

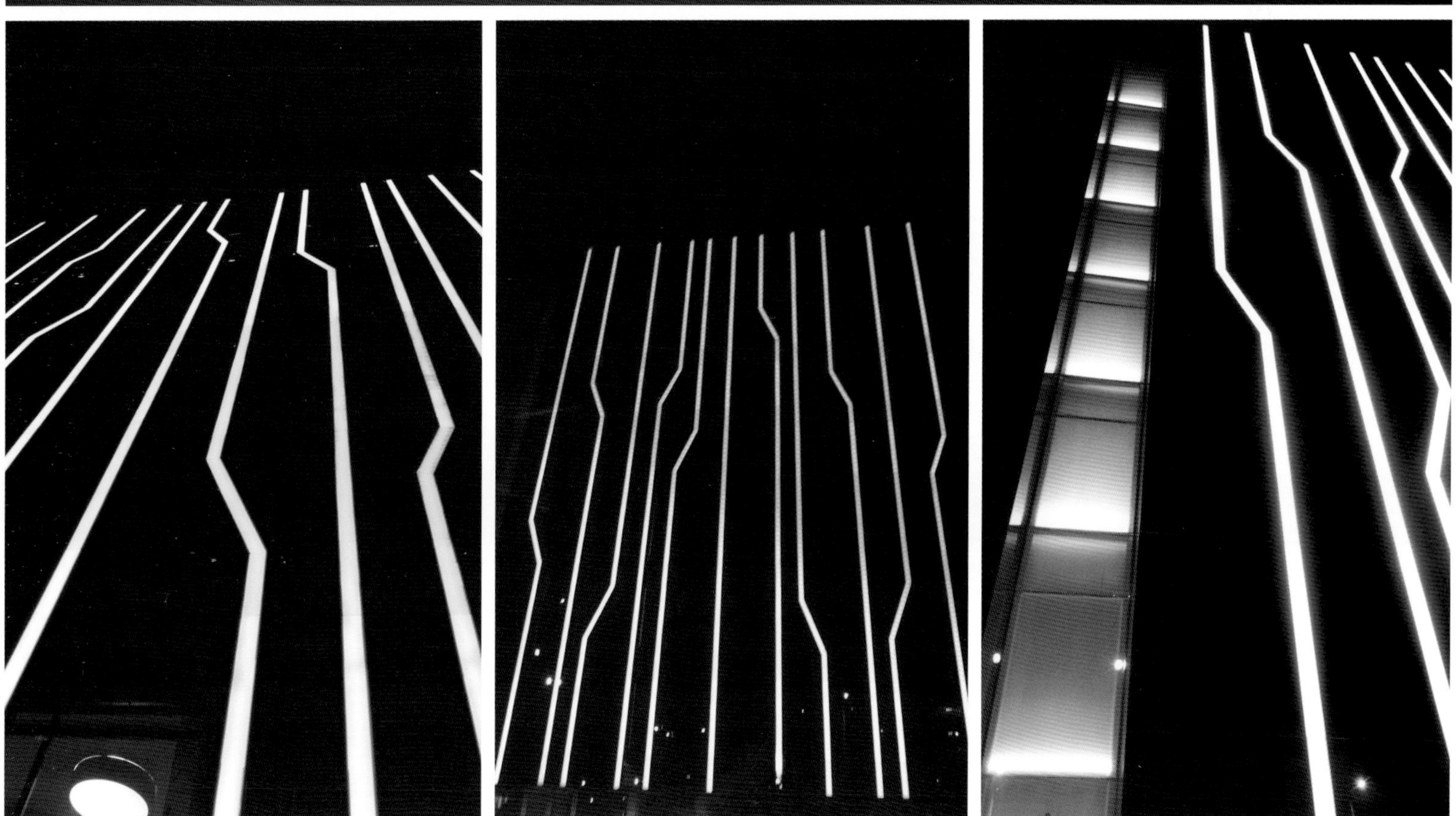

条形建筑

科研建筑

Qingpu Environmental Monitoring Station

上海青浦环境监测站

Design Company / 设计事务所:
刘宇扬建筑事务所
Design Team / 设计团队:
刘宇扬，黄昊，乐康，Kiki Cheung
Local Design Institute / 合作设计院:
上海都市建筑设计有限公司
Location / 地点:
China （中国）
Area / 面积:
5 000 m^2
Photography / 摄影:
Jeremy San

视觉亮点

建筑结构井然有序，大面积镜面材料的应用使建筑具有良好的采光条件。

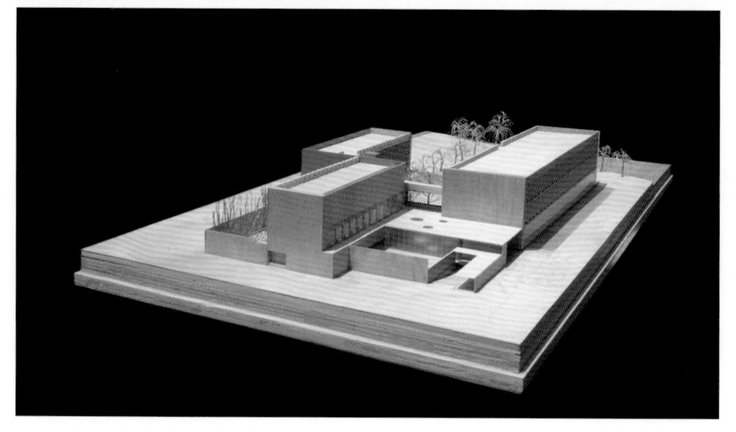

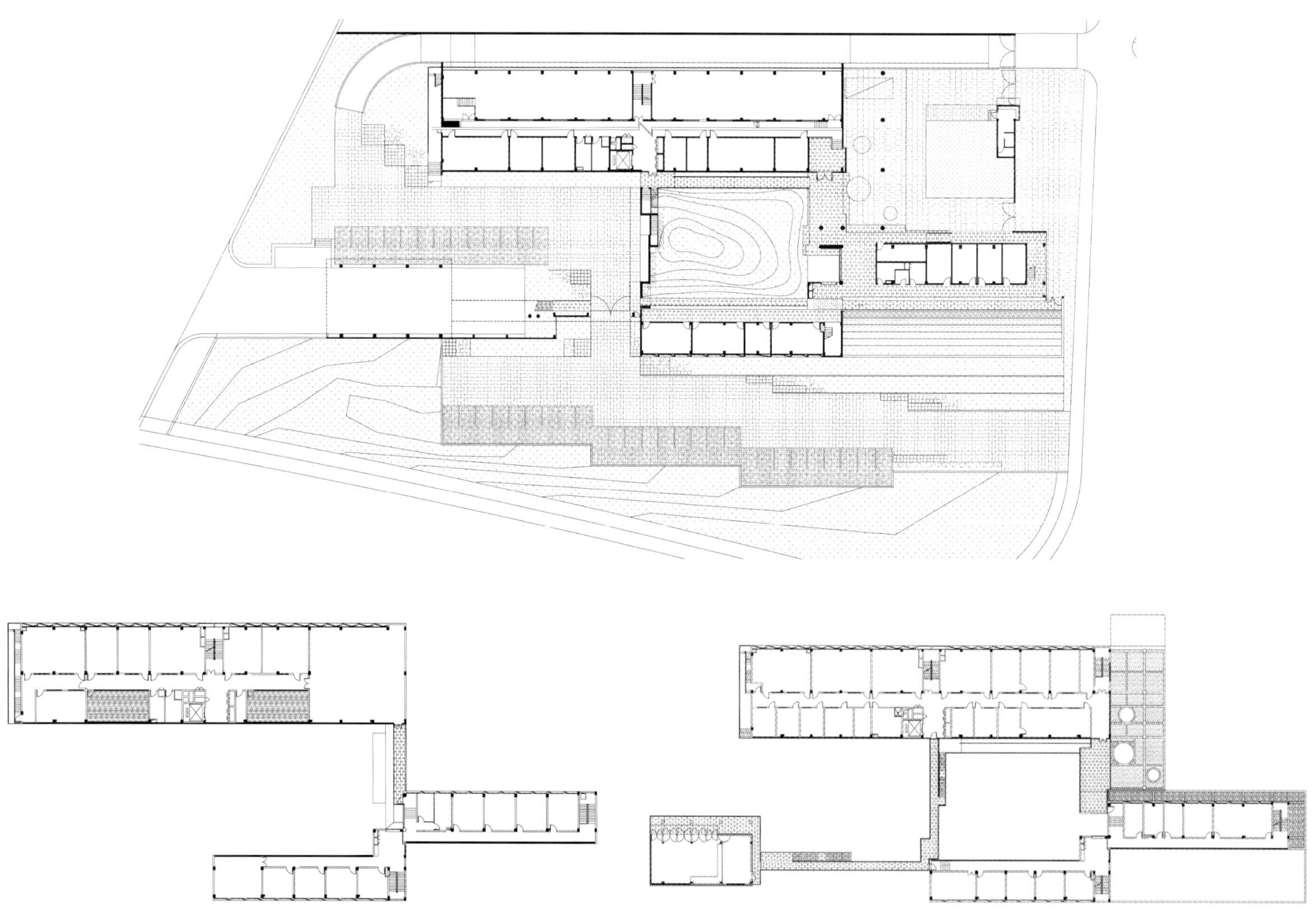

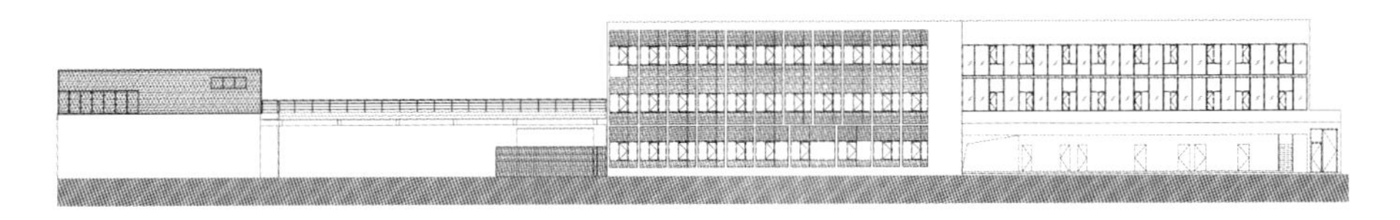

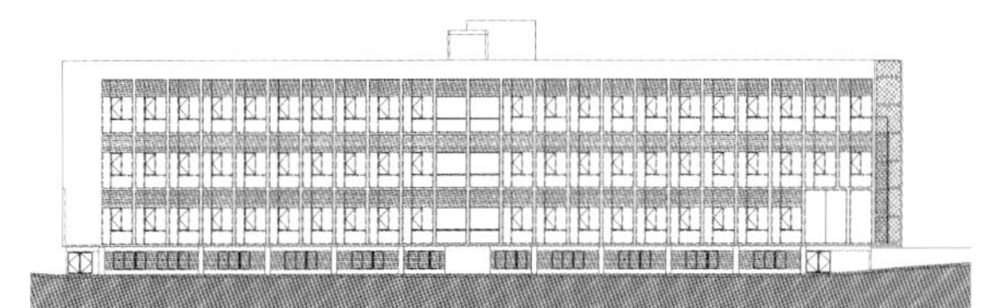

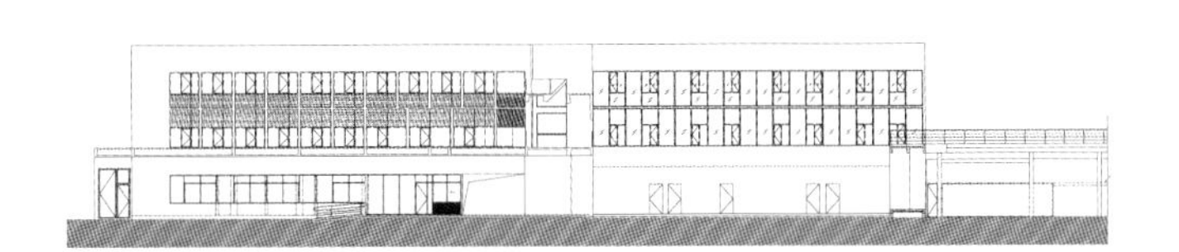

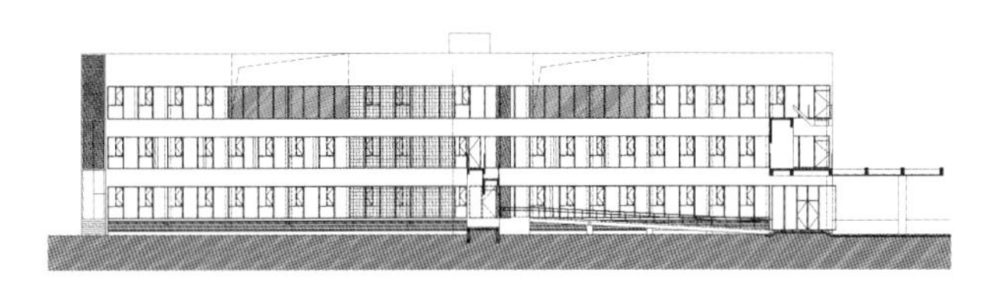

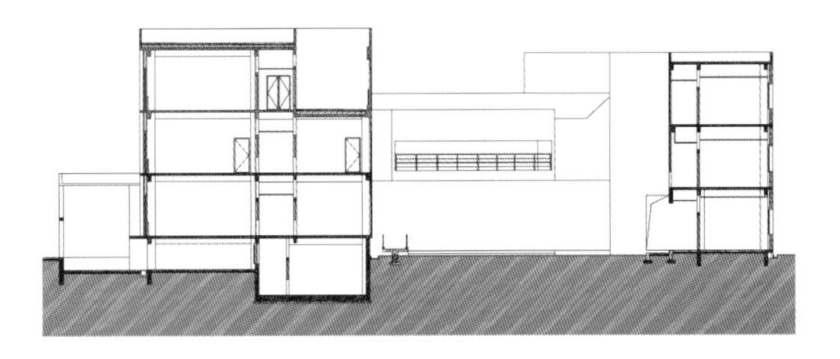

This design combines the Western "scientific view of nature" and the ancient Chinese, "humanistic view of nature" and through an individual's own observation of nature, intuitive experience of surrounding environment,forming an "phenomenal view of nature" .The design not only satisfies criteria of scientific functional layout and humanistic nature of space and form, but also utilize the visual moment when one situate in architecture to constitute the inherent relationship between man and natural environment. The design adopted the "three walls, three courtyards, the third floor," approach to blend the space in between wall and courtyard, and formed the entity of the architecture. Su Shi, "Wang Jiangnan • Spring" explains the root of Jiangnan cultural - the warm climate and detailed Xu style environment.This proposal is not focusing on the style of form of architecture, but rather, the relation between architecture and landscape (gardens, natural), and the visual moment of transitional space between buildings.

本案设计融合了西方的“科学自然观”和我国的“人文自然观”，并透过人们对自然的观察、直观式的体验，形成一种感染人们内心的“现象自然观”。在满足功能布局的科学性和空间形式的人文性之后，本案利用建筑所捕捉到的视觉瞬间，构成人与自然环境的内在关系。该建筑采取了“三墙、三院、三楼”的手法，糅合了墙与院的空间，并形成了建筑现有的形体。苏轼《望江南·暮春》中的“春未老，风细柳斜斜”一语道出了江南文化的根源——温暖的气候和优美的环境。本案最终的着墨之处不在于建筑本身的形式，而在于建筑与景观（园林、自然）的关系，以及建筑之间的过渡。

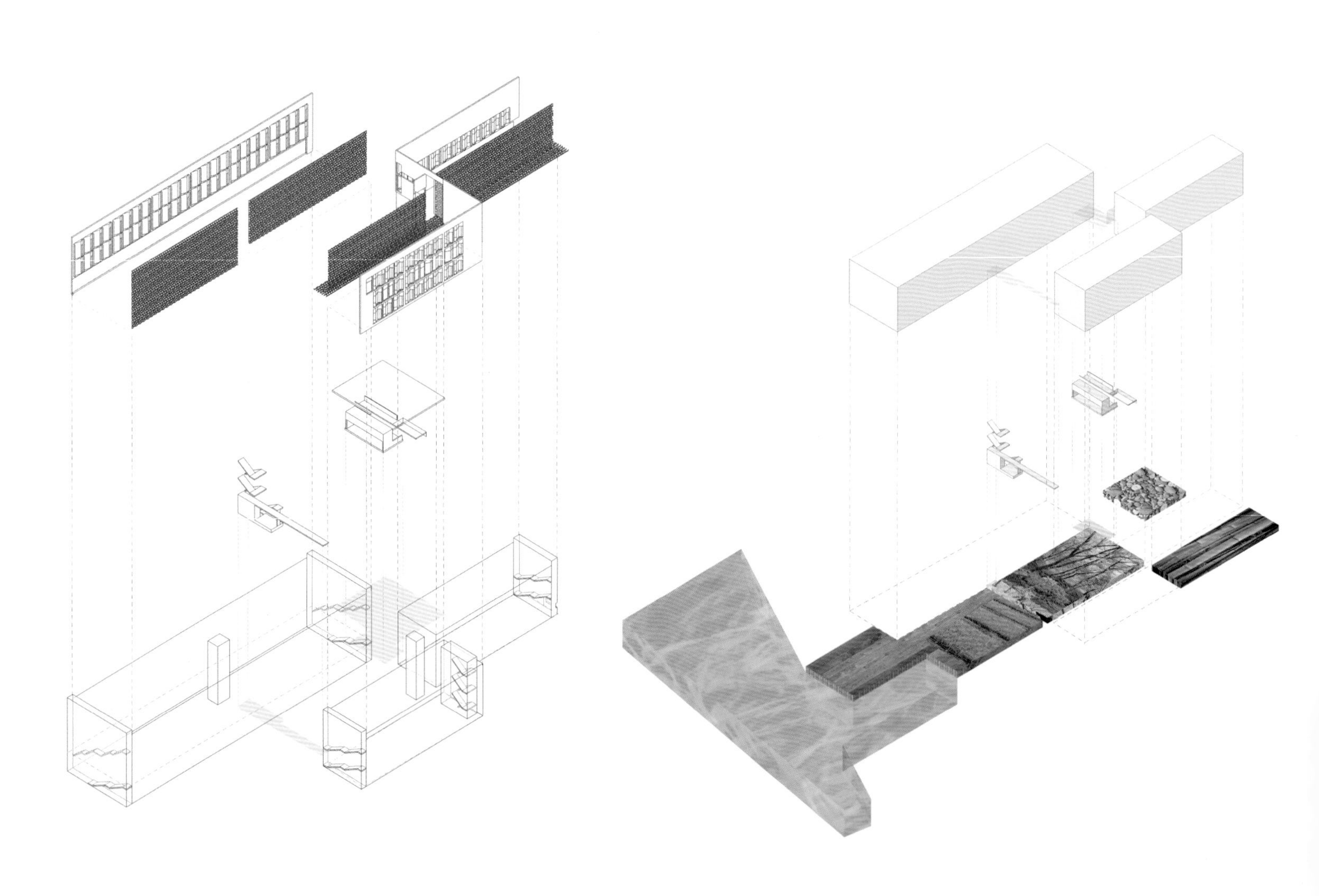

条形建筑

教育建筑

USJ Campus de L'Innovation et du Sport

贝鲁特的USJ校园

Design Company / 设计事务所:
109 Architects in collaboration with Youssef Tohme
Design Team / 设计团队:
Ibrahim Berberi, Michel Georr, Nada Assaf, Rani Boustani, Etienne Nassar, Emile Khayat, Naja Chidiac, Richard Kassab
Location / 地点:
Lebanon （黎巴嫩）
Area / 面积:
55 000 m^2
Photography / 摄影:
109 Architectes, Albert Saikaly

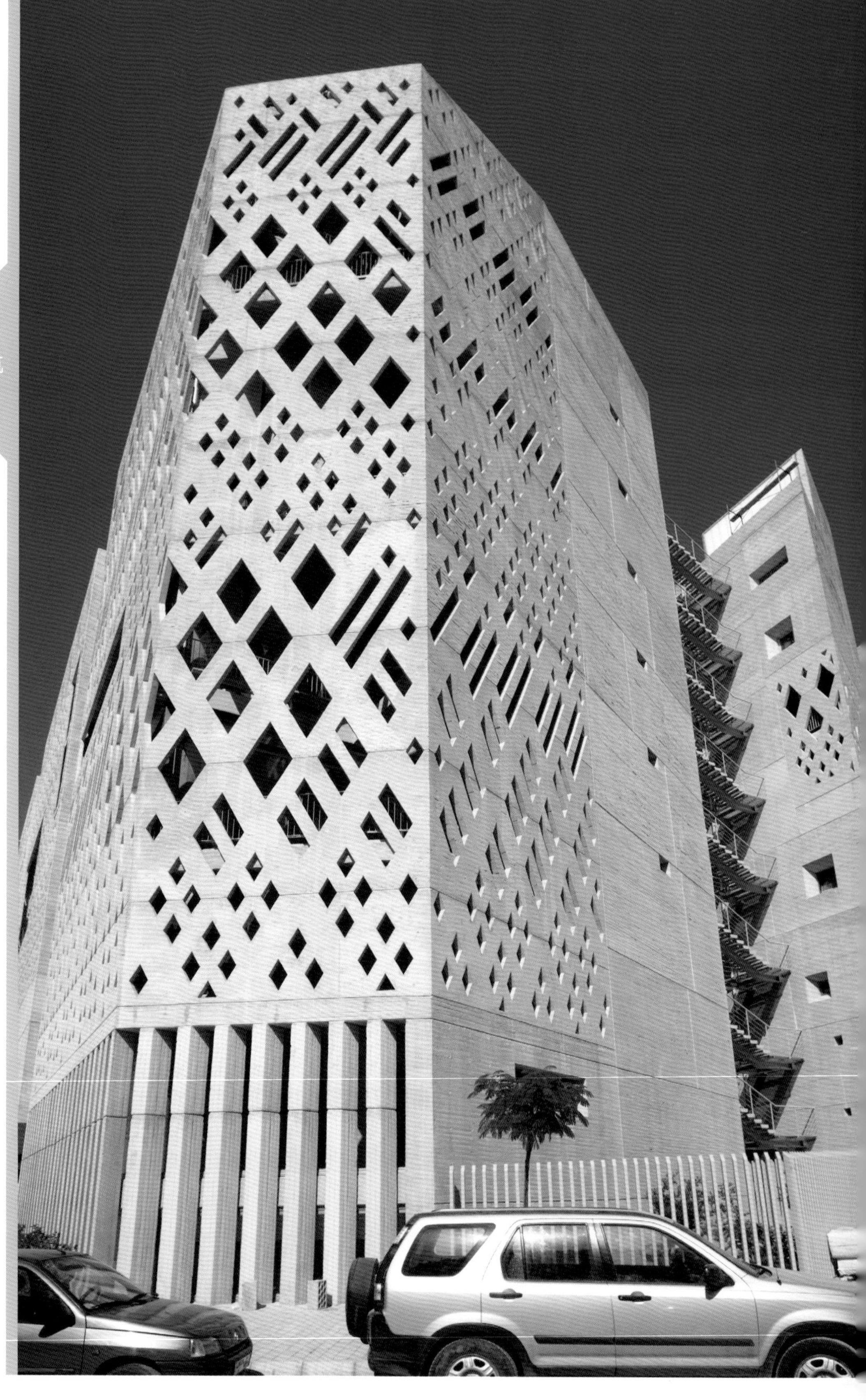

视觉亮点

建筑立面多个长方形开口设计不仅为内部空间提供了良好的采光，更赋予建筑雕塑感。

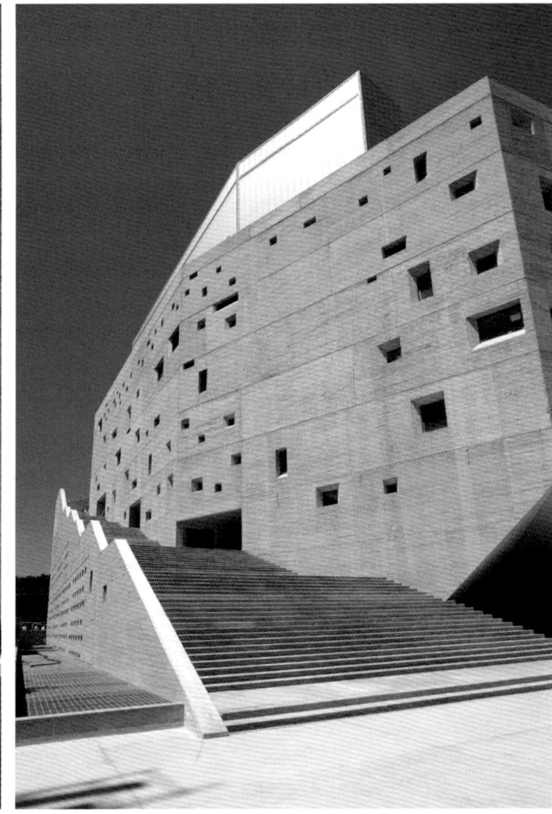

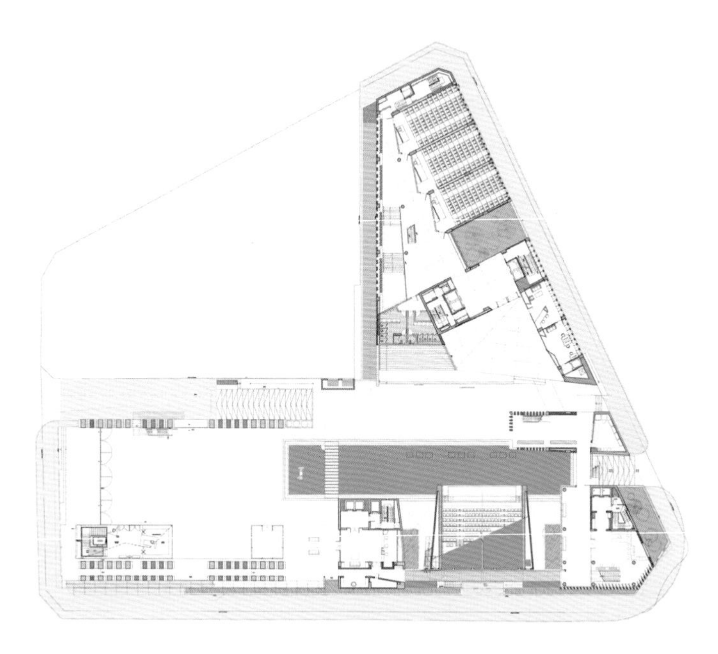

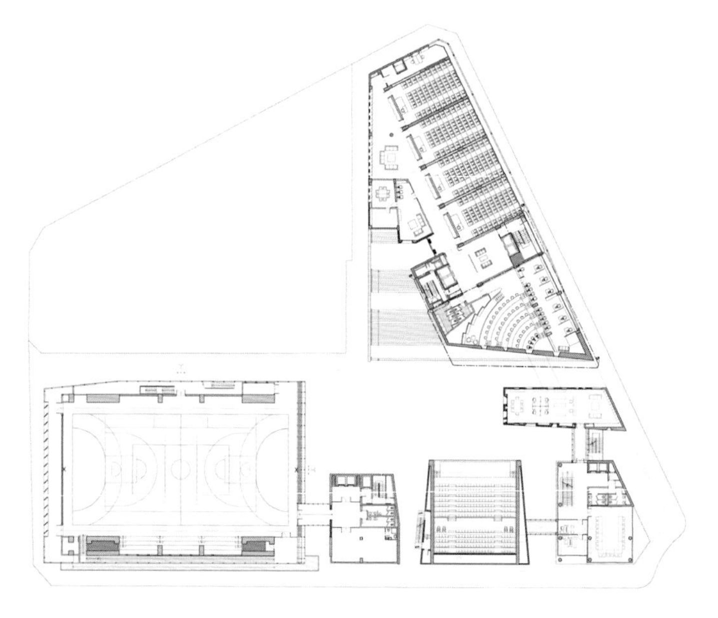

Culturally, and historically with Beirut's urban tissue. Conceptually an urban block with sculpted voids, the building's hollow spaces define six autonomous blocks and construct multiple viewpoints across Beirut, connecting students to their dynamic setting. The voids also generate a street-level meeting space, which flows fluidly to the top floor in the form of a massive staircase. It concludes at a landscaped terrace overlooking the city. Light is a vital element in oriental architecture and one that shapes its style and identity; the campus exposes alternate light qualities through Moucharabieh-inspired perforations and a polycarbonate volume. Such manipulation presents a striking contrast in filtered light and luminescence.

设计师力求新建的建筑能够体现贝鲁特城市的文化和历史。从建筑外观来看，这就是一座极具雕塑感的城市建筑。建筑内共有六处相互独立的空间，在这里可以获得欣赏贝鲁特城市风景的多重视角，进一步加强了学生们与富有活力的城市环境之间的互动。这座建筑从一个城市街区中突显出来，成为一个集会空间。建筑师在室外设置了一个巨大的楼梯，从垂直方向上引导学生们进入建筑内部。楼梯的尽头连接了一个景观平台，在这里学生们可以观看到全城的景色。该建筑通过孔状结构实现室内自然采光。

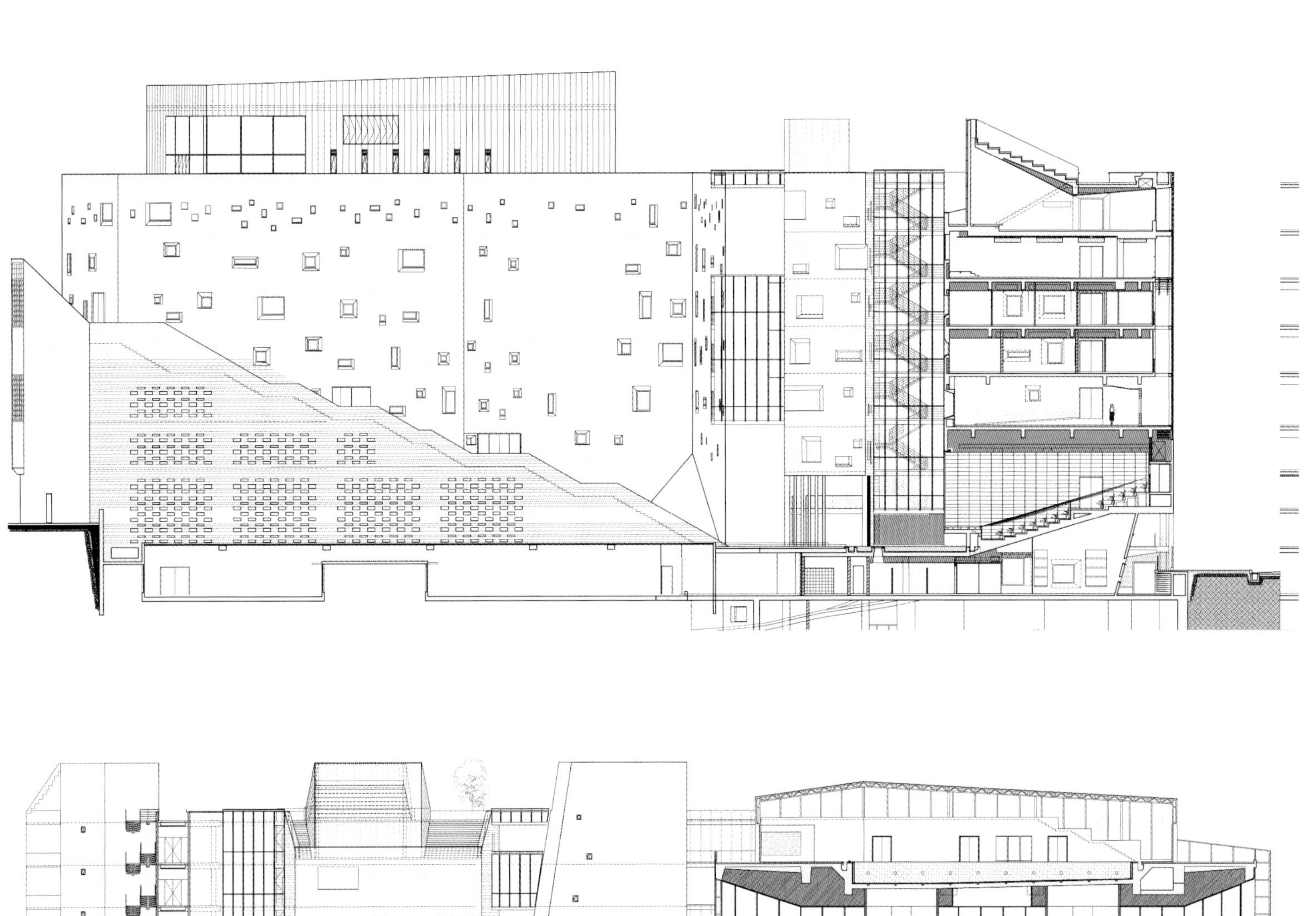

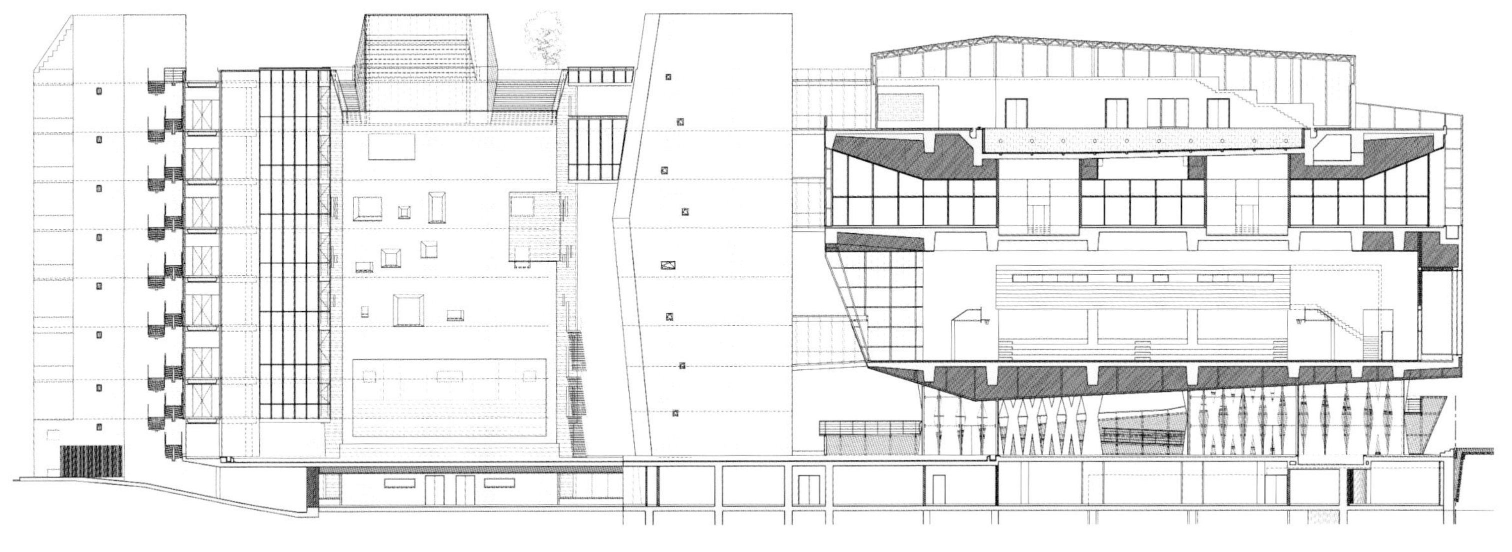

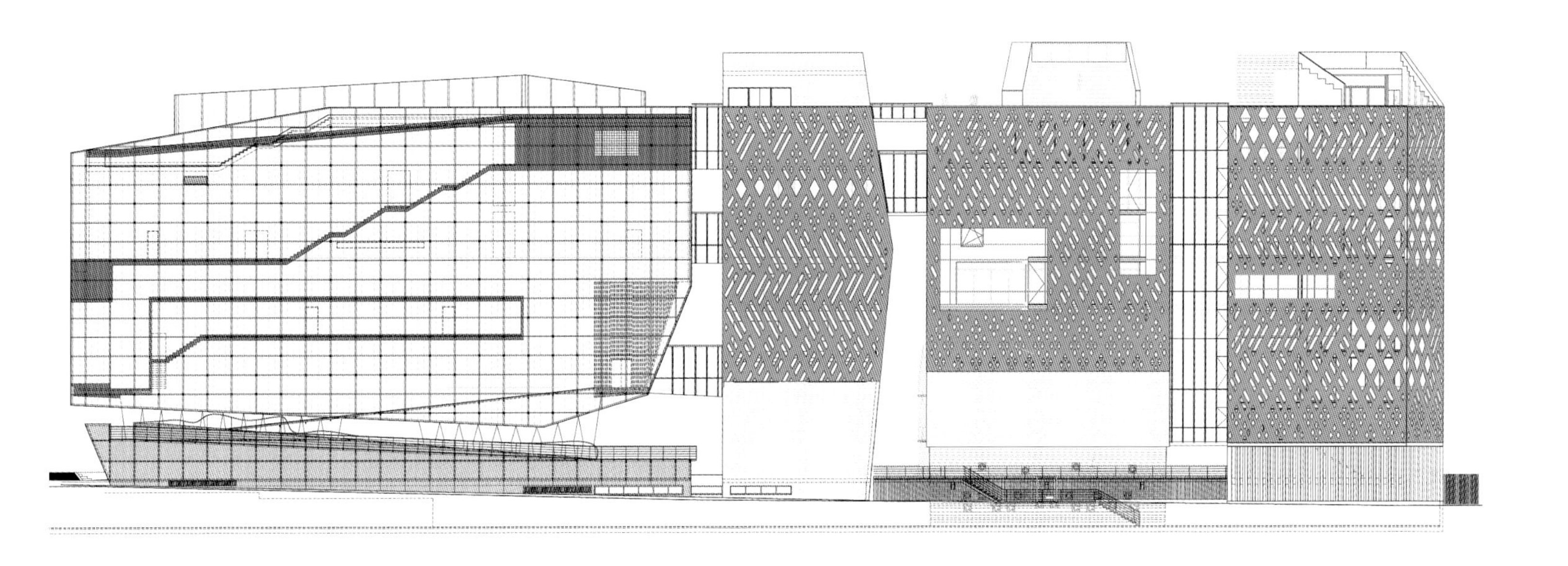

条形建筑

医疗建筑

Hospital Klinikum Klagenfurt

克拉根福医院

Design Company / 设计事务所:
Dietmar Feichtinger Architectes

Location / 地点:
Austria （奥地利）

Area / 面积:
site 236 590 m², floor 75 000 m²

Photography / 摄影:
Wolfgang Thaler, Hertha Hurnaus, Gisela Erlacher, Foto Horst

视觉亮点

建筑立面整齐统一，建筑结构极具秩序感，建筑间的庭院空间为人们创造了一个舒适而优雅的就医环境。

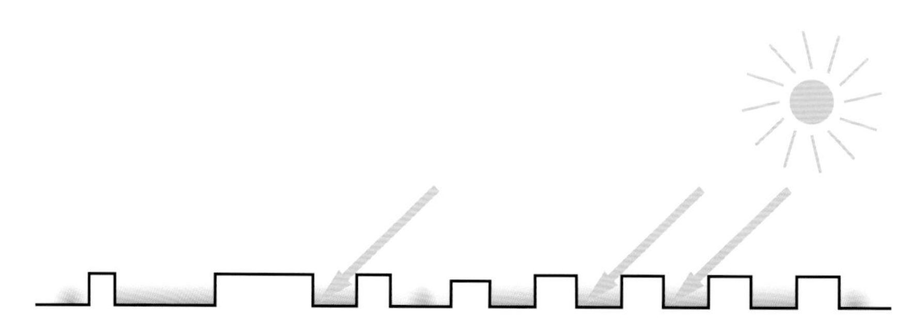

Feschnig Straße
Glan
ALLGEMEINPFLEGE
TAGESPFLEGE
IMC
INTENSIVPFLEGE
DIAGNOSE
OP
DIENSTLEISTUNGEN
PERS. GARDEROBE
BÜRO
ANLIEFERUNG / TGM
ENTSORGUNG
WÄSCHEREI
KÜCHE
APOTHEKE
STERILISATION
LABOR
ERSCHLIESSUNG
TECHNIK
PARKEN

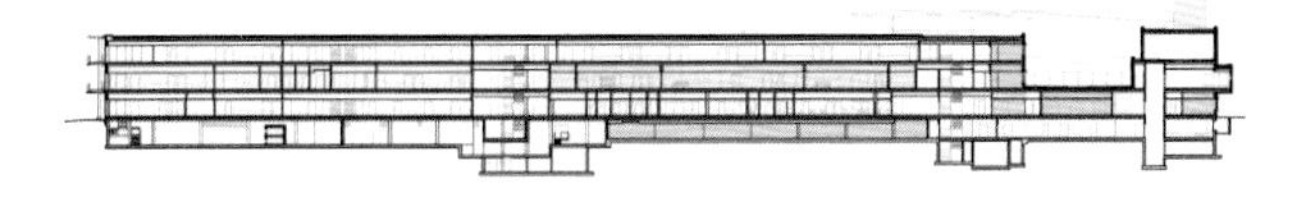

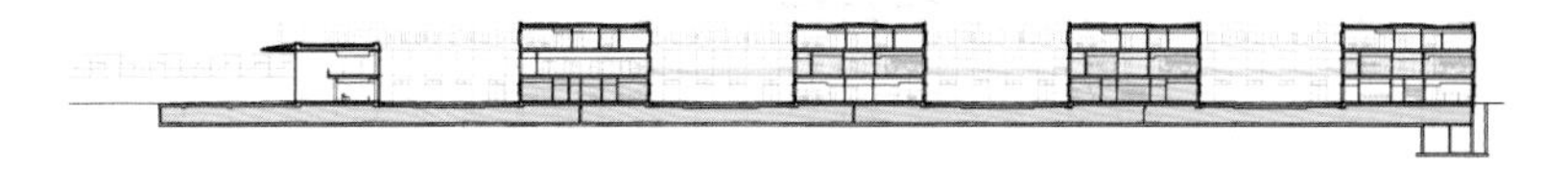

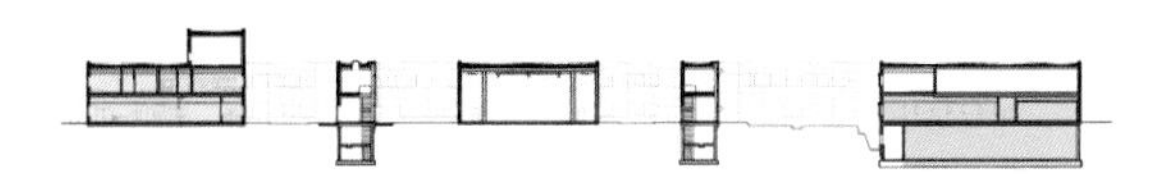

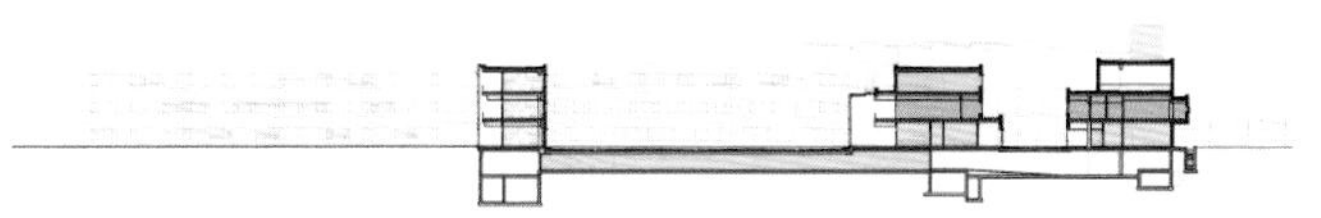

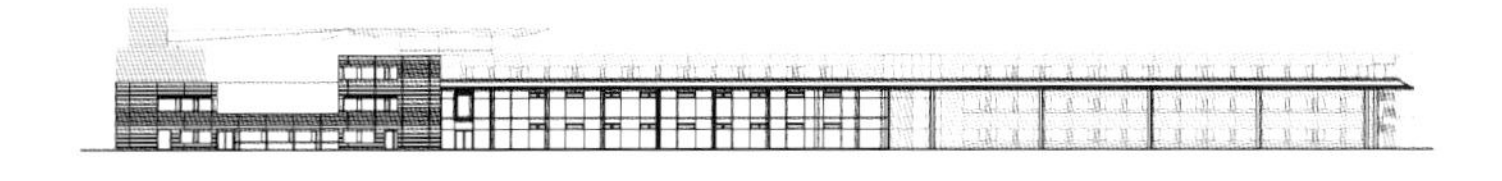

EINGANGSHALLE	*FOYER*	DIAGNOSE	*DIAGNOSTICS*	ANLIEFERUNG /TGM	*LOGISTICS / TECHN. FACILITY MANAGEMENT*	STERILISATION	*STERILIZATION*
ALLGEMEINPFLEGE	*IN-PATIENTS*	OP	*SURGERY*	ENTSORGUNG	*DISPOSAL*	LABOR	*LABORATORIES*
TAGESPFLEGE	*DAY CARE*	DIENSTLEISTUNGEN	*SERVICES*	WÄSCHEREI	*LAUNDRY*	ERSCHLIESSUNG	*COMMUNICATION SERVICES*
IMC	*INTERMEDIATE CARE*	PERS. GARDEROBE	*CLOAKROOM STAFF*	KÜCHE	*KITCHEN*	TECHNIK	*TECHNOLOGY*
INTENSIVPFLEGE	*INTENSIVE CARE*	BÜRO	*OFFICES*	APOTHEKE	*PHARMACY*	PARKEN	*PARKING*

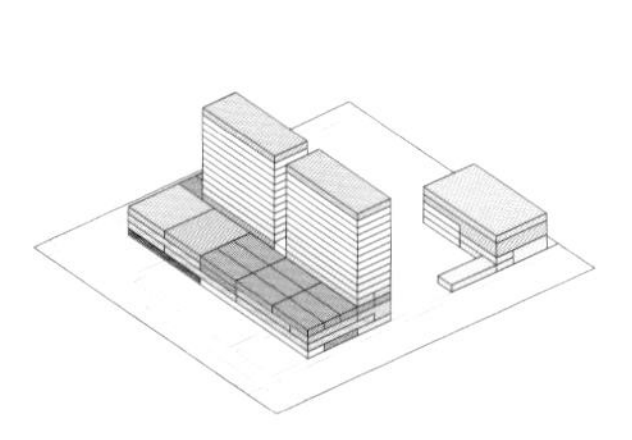

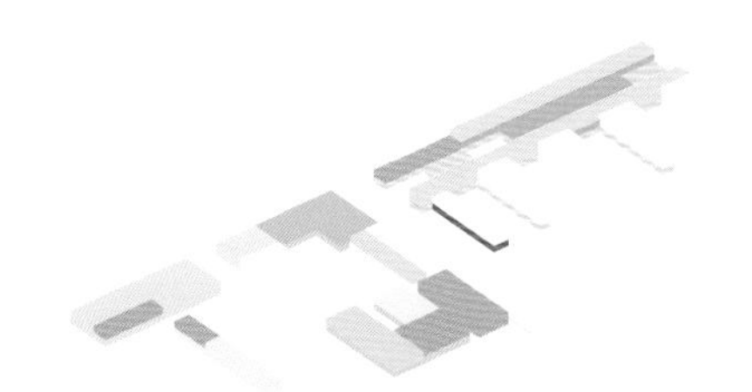

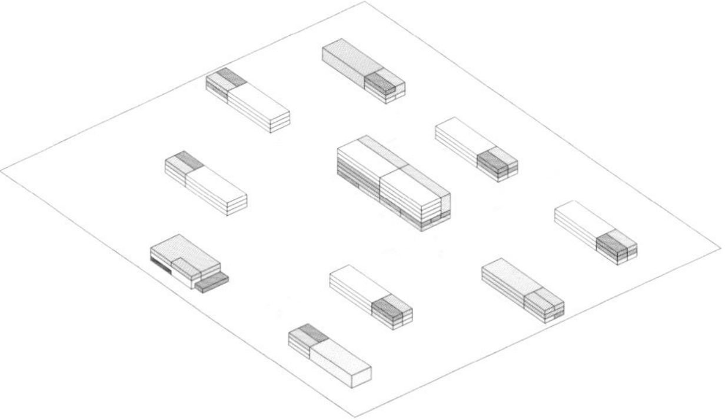

A modern, innovative and trend-setting hospital

The new project for the Provincial Hospital in Klagenfurt is innovative in many ways. The latest MedicalTechnology along with an enhanced cross-utilization of medical equipment and facilities (operatingrooms, examination and treatment rooms and wards in the logistics field) give this new hospital apioneering status in Europe. Even before completion, it served as an example to future developments.The architectural concept is an important part of this modern definition of a hospital. The flat, two-storeybuilding fits well into the landscape on a site where the wetlands of the Glan River close off the land tothe north. The footprint of the building is largely determined by landscaped courtyards which open thebuilding up to the site as well as carefully form more private areas for its users. The horizontality of thebuilding is reinforced by the two main access routes: the curved corridor to the north, and the straightcorridor which accesses the examination and treatment areas. Large multi-storey glass facades givethese zones, which are designated as the waiting and circulation areas, an inviting and more transparentopen character.

The horizontal division of the façade and the tiered structure of the form combine tofurther integrate the complex into the landscape.

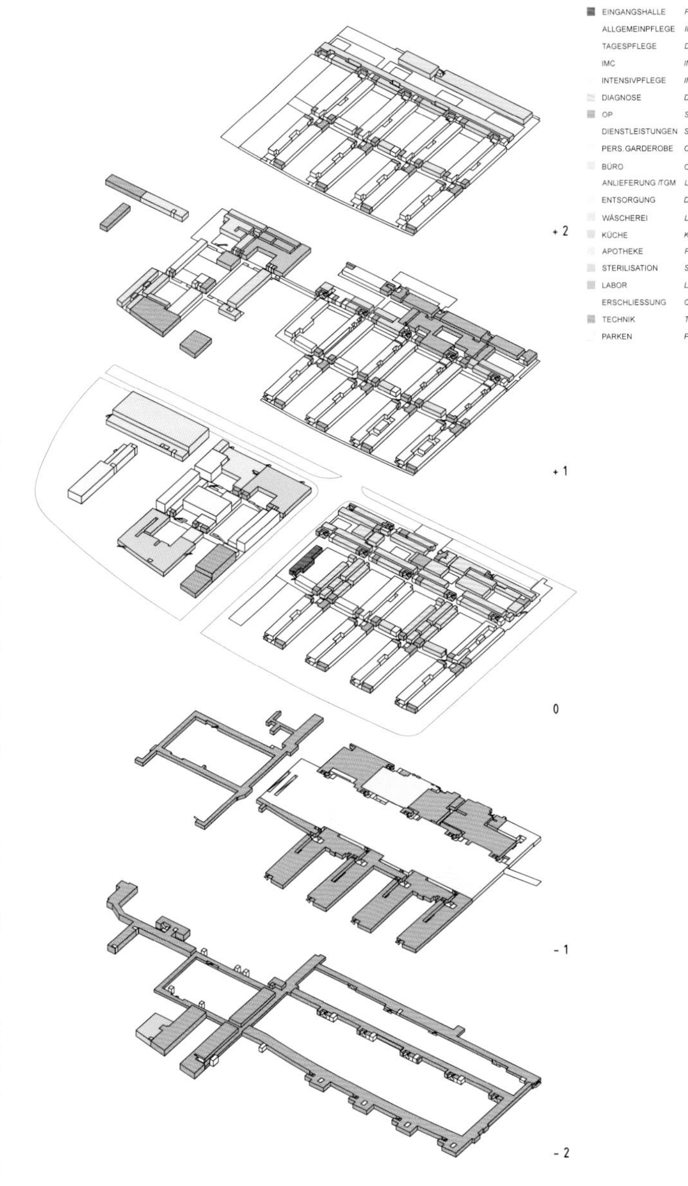

位于奥地利克拉根福的这所新建的医院在很多方面都极具创新性。医院拥有最先进的医疗技术，建筑空间方便医护人员交叉使用各种医疗设备，所有这些都使这座医院在欧洲处于领先地位。这座外观平整流畅的两层建筑与周边景观环境完美相融，所处地块的北部为格兰河湿地。该建筑所处的地块基本由一处景观庭院来界定，该庭院将建筑与整个地块联系在一起，同时为人们提供了一些较为私密的空间。两条建筑主通道突出了整座建筑的水平式空间设计特色：一条为建筑北侧的曲折式走廊，一条为可通向体检区和治疗区的笔直通道。大型的多层式玻璃立面赋予该建筑以通透、开放式的特色。

建筑立面的水平式划分与建筑的分层结构进一步将整座建筑与周边景观环境融为一体。

条形建筑

教育建筑

Binh Duong School

越南平阳初中

Design Company / 设计事务所:
Vo Trong Nghia Co., Ltd.
Project Architect / 设计师:
Vo Trong Nghia, Shunri Nishizawa, Daisuke Sanuki
Location / 地点:
Vietnam （越南）
Area / 面积:
floor 6564 m^2
Photography / 摄影:
Hiroyuki Oki

视觉亮点

建筑立面设置连续的条形混凝土百叶窗，不仅遮风挡雨，还营造出笼子般的建筑表皮。

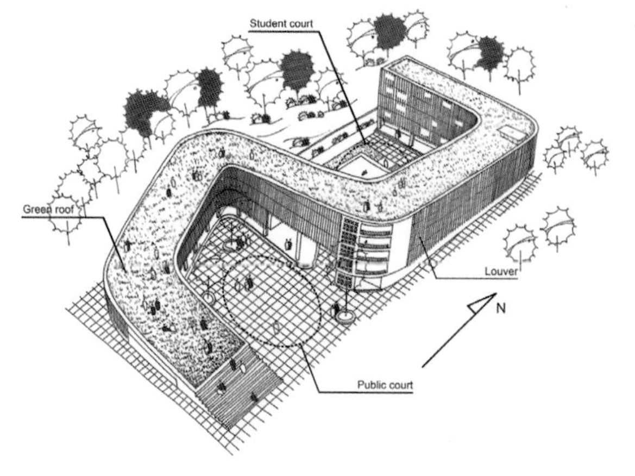

A
A
1
2
3
4
5
6
7
8
9
10
11
12
13
14
15
16
17
1. Main entrance
2. Protect room
3. Garage for teacher
4. Entrance to roof
5. Multipurpose hall
6. Office
7. Meeting room
8.Reading rdm for pupils
9.Classroom
10.Kitchen
11. Canteen
12.Terrace
13. Swimming
14. Engineering Department
15.Public court
16.Student court
17.Common space
18.Green roofing

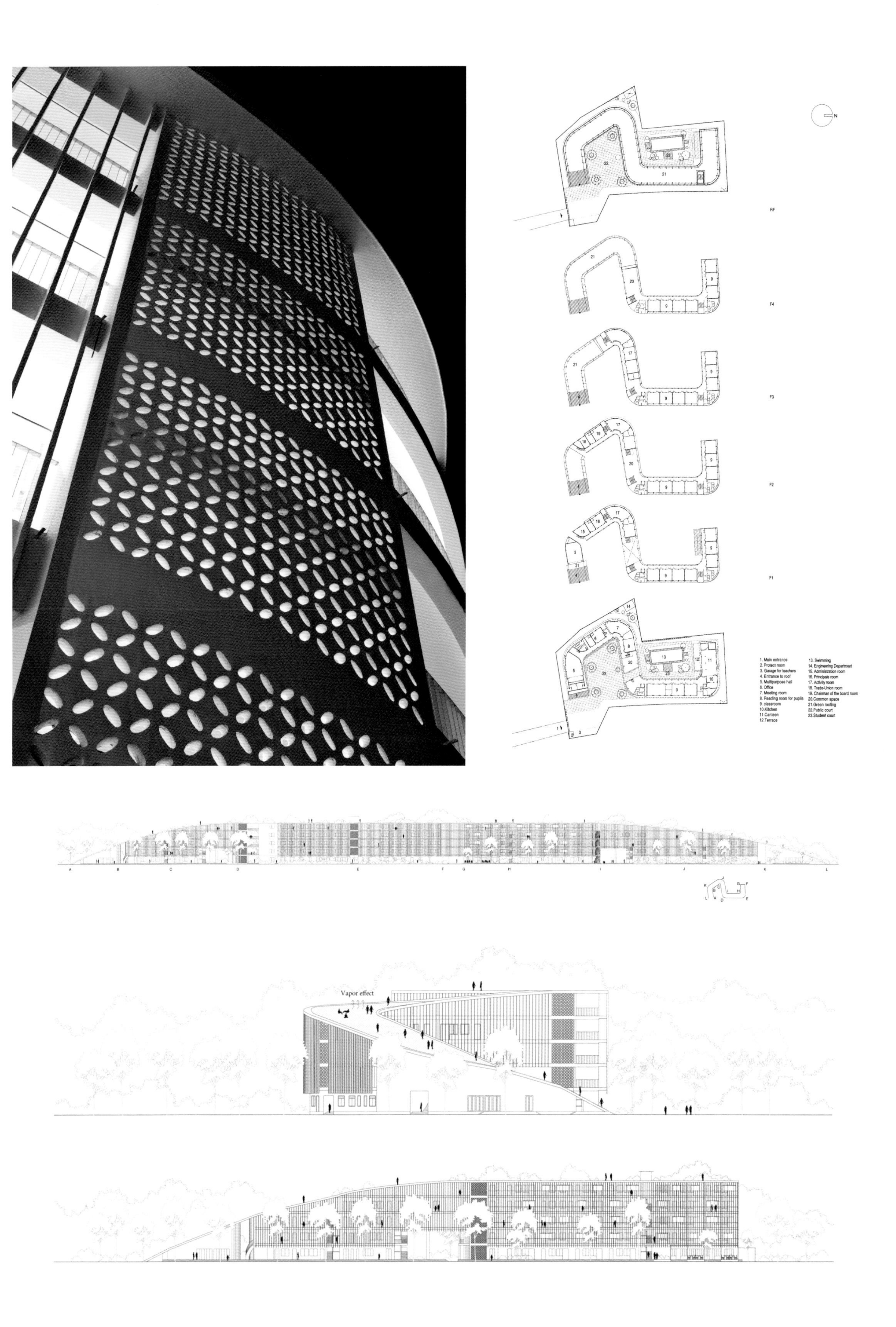
RF
F4
F3
F2
F1
1. Main entrance
2. Protect room
3. Garage for teachers
4. Entrance to roof
5. Multipurpose hall
6. Office
7. Meeting room
8. Reading room for pupils
9. classroom
10.Kitchen
11.Canteen
12.Terrace
13. Swimming
14. Engineering Department
15. Administration room
16. Principals room
17. Activity room
18. Trade-Union room
19. Chairman of the board room
20.Common space
21.Green roofing
22.Public court
23.Student court
Vapor effect

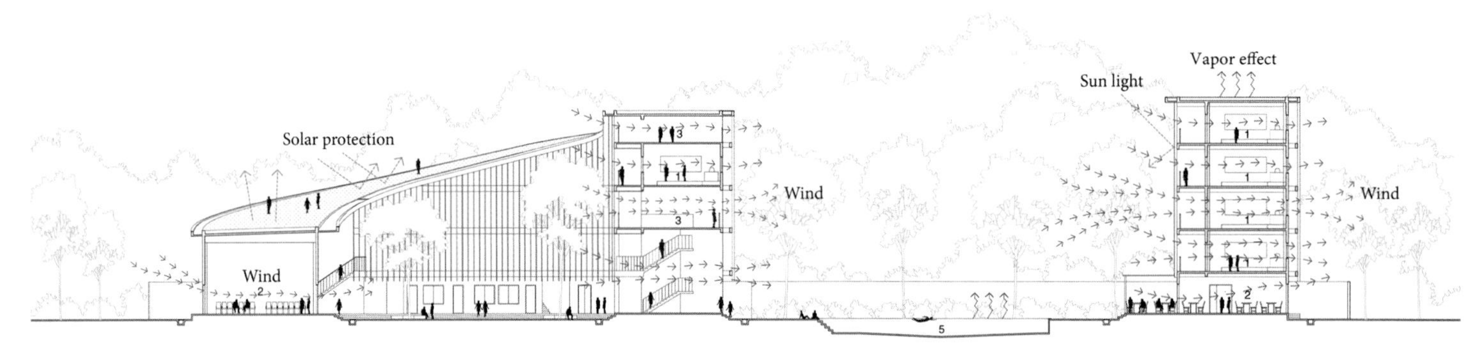

Binh Duong, a new city which is 30 minutes away from Ho Chi Minh City, has a typical tropical climate all year round. The site is located in the middle of a flourishing forest with a wide variety of green and fruits, running rampant.

The building is located in 5300 square meters abundant land, consisting of a maximum height of five levels, with the intention of being surrounded by the height of the forest around. Pre cast concrete louvers and pattern walls are used for envelop of the building. These shading devices generate semi-outside space, these open circumstances avoiding direct sunlight as well as acting like a natural ventilation system for the corridor space. All the classrooms are connected by this semi-open space, where teachers and students chatting, communicating and appreciating nature. We designed the school as a continuous volume in order not to disturb any school activities. This fluidity concept is inspired by the endless raining of the typical tropical climate, where raining season lasts from May to November each year.

新城平阳距离胡志明市只有 30 分钟的车程，终年均为典型的热带气候。该项目地块位于一片茂密森林的中央位置，这里有种类繁多的绿色观赏植物和果树。该项目占地面积为 5300m²，最高的建筑为五层。较低的高度设计是为了保证整座学校都处于森林中高大树木的庇护中。预制混凝土百叶窗和混凝土墙体包裹着整座建筑。这些遮蔽设施不仅避免阳光直射进建筑内部，还成为走廊区的自然通风系统。所有教室都通过这些半开放式的空间结构联系在一起，师生们可以在这里闲谈、交流、欣赏美妙的大自然。这种建筑结构设计的灵感来自于热带地区连绵不断的雨，这里的雨季能从每年的 5 月一直持续到 11 月。

条形建筑

办公建筑

Head Office Fiteco

法国Fiteco集团办公楼

Design Company / 设计事务所:
Colboc Franzen & Associés
Location / 地点:
France（法国）
Area / 面积:
7 005 m^2
Photography / 摄影:
Cécile Septet

视觉亮点

镜面材料和金属材料完美搭配，创造了一个不同寻常的办公建筑。

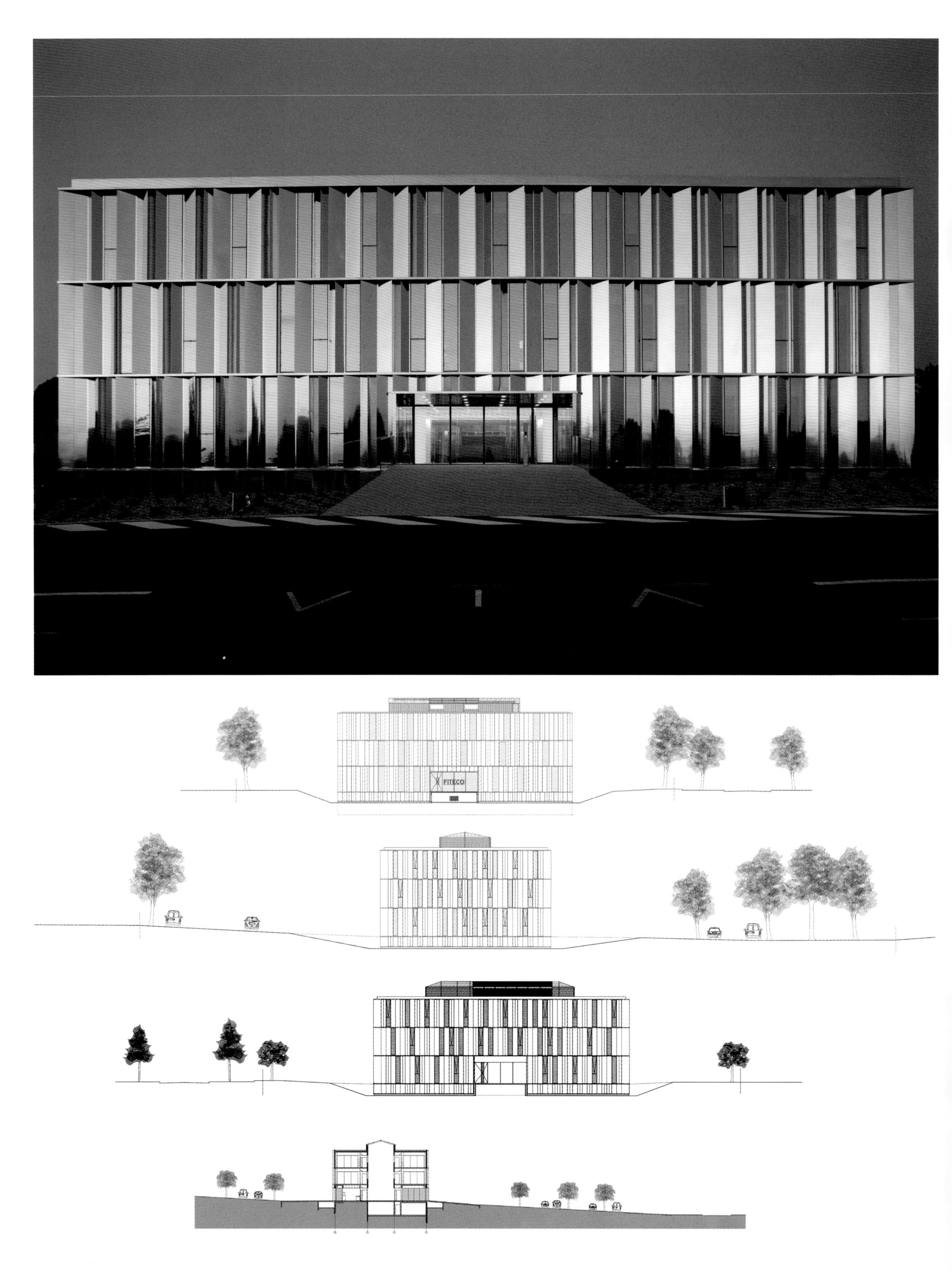
FITECO

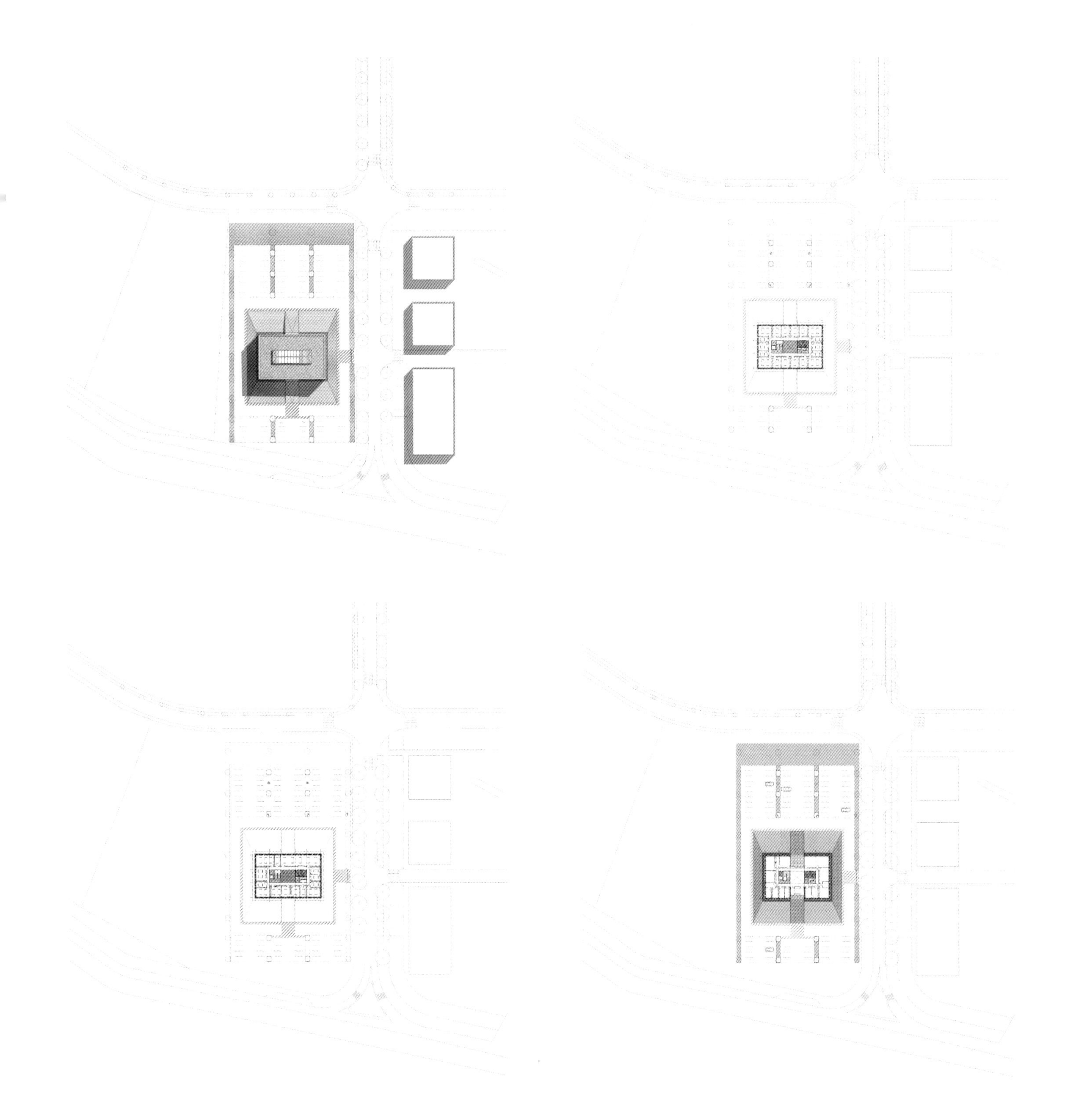

The French architecture office Colboc Franzen & Associés decided to go further and to design a single building, so the construction would cost less, the environmental impact of the building would be reduced and the internal organization would be optimized. On the first floor are situated all the common facilities. The local branch settled down on the second floor and the head office on the third and last floor.
The heart of the building is composed of an atrium, two cores of services (staircase, elevators, restrooms, photocopy room...). Interior is kept in white, floors in light grey; only the setting of the ceiling lights creates a noticeable graphisme. The air conditioning system running all around each floor at the ceiling along the façade, a uniform positioning of the power sockets, and the movable dividing (non load-bearing) walls give its versatility to the building. Therefore the neutrality of the working environment doesn't interfere with the worker's concentration.

承接该项目的法国建筑事务所决定打造一座单独的建筑，并控制建筑成本，减小建筑对周边环境的影响，同时最大限度地优化内部空间设置。所有的公共设施均位于建筑一层，本地分支机构位于建筑二层，总部办公室位于建筑三层。建筑的中心部分设置了中庭和两处核心服务区（楼梯、电梯、休息室、影印室等）。内部空间色调为白色，地板为浅灰色，顶棚照明灯的背景为醒目的色彩。每个楼层的空调系统均沿立面一侧的顶棚设置，其中包括统一设置的电源插座、可移动式分隔墙体（非承重墙）等，这些均提升了整座建筑的功能性。较为中性化的室内工作环境不会分散作人员工作时的注意力。

条形建筑

市政建筑

House of Justice

市政大厅

Design Company / 设计事务所:
Architects of Invention
Design Team / 设计团队:
Niko Japaridze, Gogiko Sakvarelidze, Dato Canava, Eka Kankava, Eka Rekhviashvili, Viliana Guliashvili, Nika Maisuradze, David Dolidze, Soso Eliava, PM Devi Kituashvili
Location / 地点:
Georgia（格鲁吉亚）
Size / 面积:
site 1 100 m^2, building 3 638 m^2, total floor 1 438 m^2
Photography / 摄影:
Nakanimamasakhlisi Photo Lab

视觉亮点

建筑由矩形体块和椭圆形体块组成，建筑结构简洁、整齐，颇有秩序感，营造了市政建筑较为严肃的氛围。

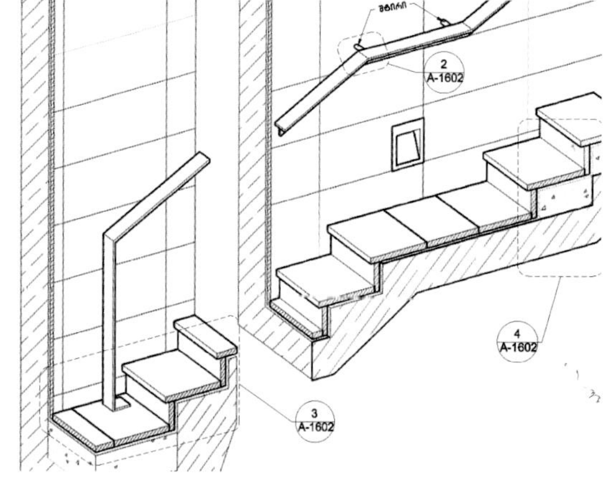

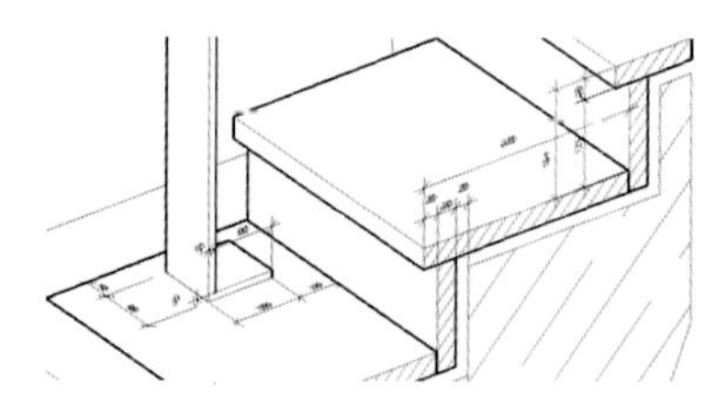

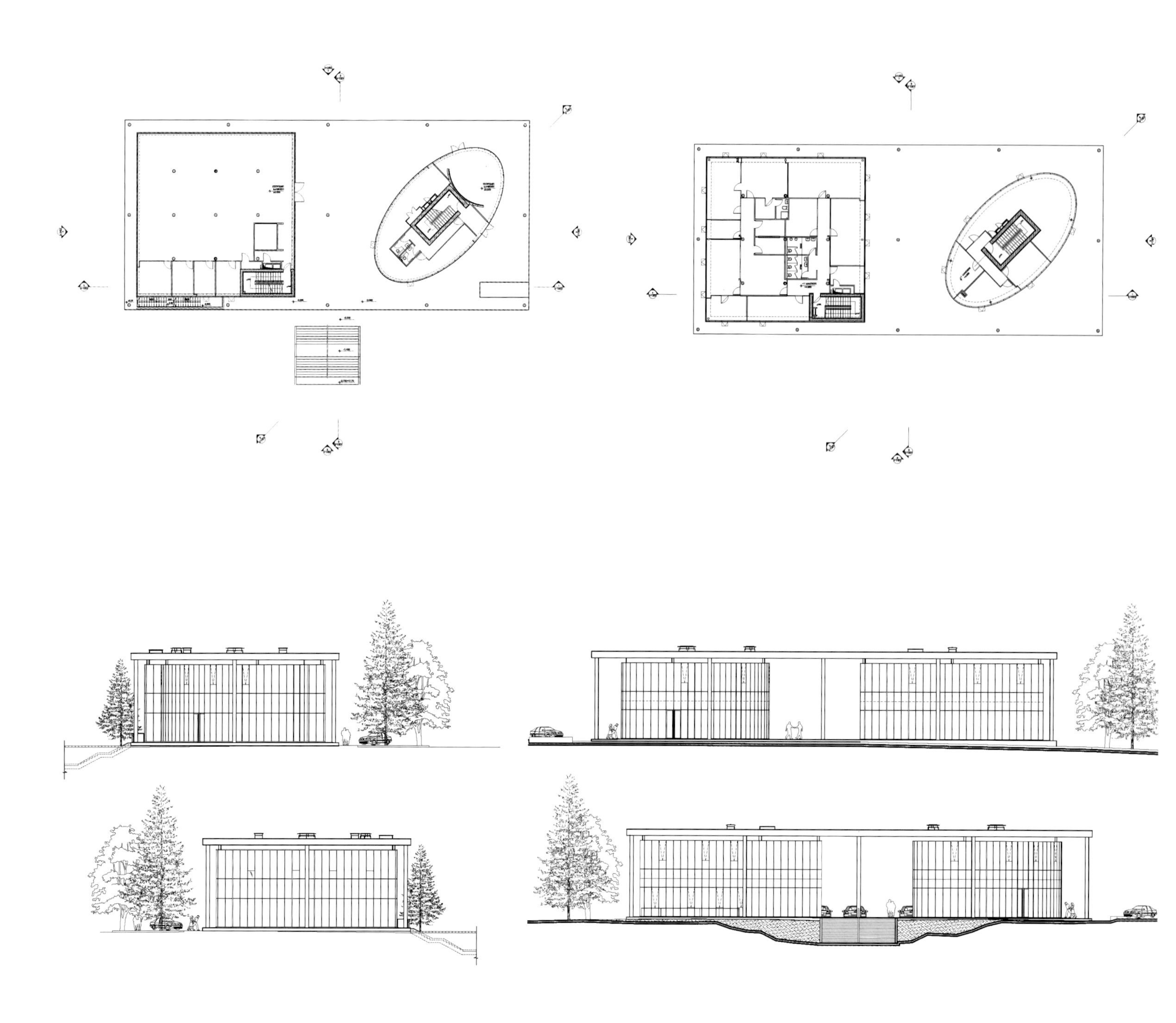

House of Justice is situated in the North-West part of Ozurgeti on Southern side of K. Gamsakhurdia square, on the adjacent site of the local museum of art. There is a four hundred seat theatre on the Eastern side of the square, residential building on the Northern and boulevard on Wester sides of the square. Building height is adjusted to the height of the museum, so it creates a symmetrical courtyard between the buildings.

The building consists of two fifty by twenty horizontal planes, connected with ten perimetrical columns and two independent glass volumes. Each volume is double stories connect wit each other with underground path. Top horizontal plane acts as a common roof for both volumes. Lower horizontal plane acts as common platform and a public space. One of the volumes is rectangular in plan, nineteen by nineteen meters and consists of - ground floor Public Service Hall and offices. Another volume is oval in plan consists of - ground floor Wedding Hall and offices. Both volumes have 100% transparent low emission glass. All the surfaces interior and exterior are composed in white tone.

建筑坐落在奥祖尔盖蒂西北部广场的南侧，毗邻当地的艺术博物馆。广场的东部是一座拥有400个座位的剧院，广场北部是住宅建筑。建筑的高度要与博物馆相协调，因此设计师在两个建筑之间设置了一处庭院。

建筑的主体结构由十根柱子支撑。建筑外立面为玻璃幕墙，每组玻璃幕墙组成的空间分为两层，同时通过通道连接。建筑巨大的屋顶覆盖了所有的空间。一层平面提供了大面积的公共空间。办公区域和服务大厅空间呈矩形，规模为19m×19m，另一组办公区域呈椭圆形。白色是室内外设计的主要色调。

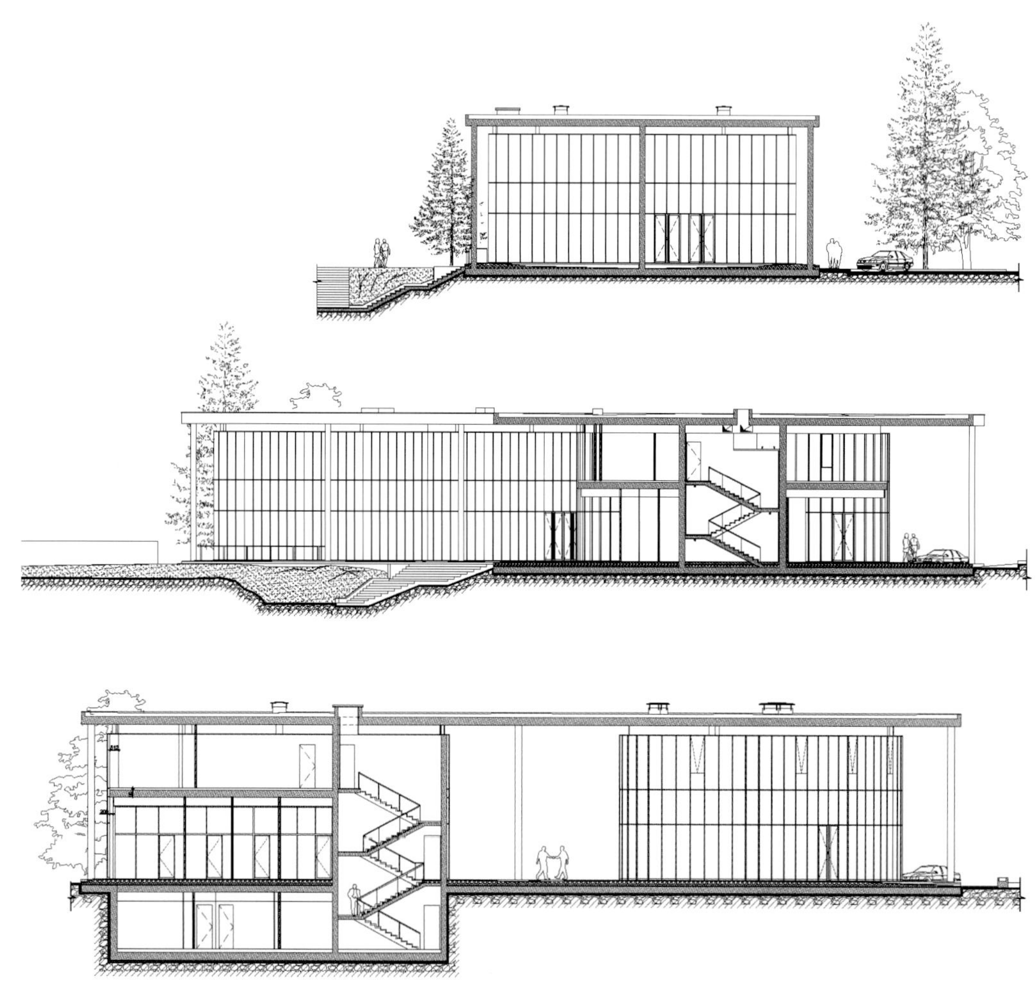

Area

条形建筑

Economic & Masters Building UNAV

经济和商业学院大楼

Design Company / 设计事务所:
Otxotorena Arquitectos
Project Architect / 设计师:
Juan M. Otxotorena, Gloria Herrera (Collaborator Architect), Catalina Delgado (Collaborator Architect), Jorge Ortega (Collaborator Architect), Isabelino Río (Collaborator Architect), Ignacio Quintana (Collaborator Architect)
Location / 地点:
Spain （西班牙）
Area / 面积:
15 529.60 m²
Photography / 摄影:
Juan Rodríguez, Pedro Pegenaute, José Manuel Cutillas, Rubén Pérez Bescós

视觉亮点

建筑外观简洁、大方，富有节奏感的片层混凝土结构是建筑最出彩的地方。

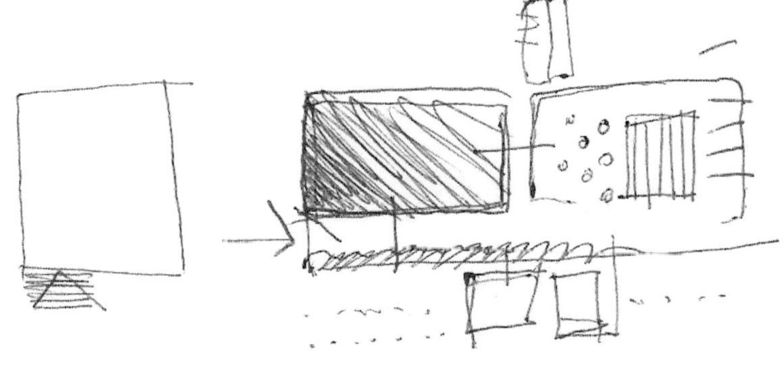

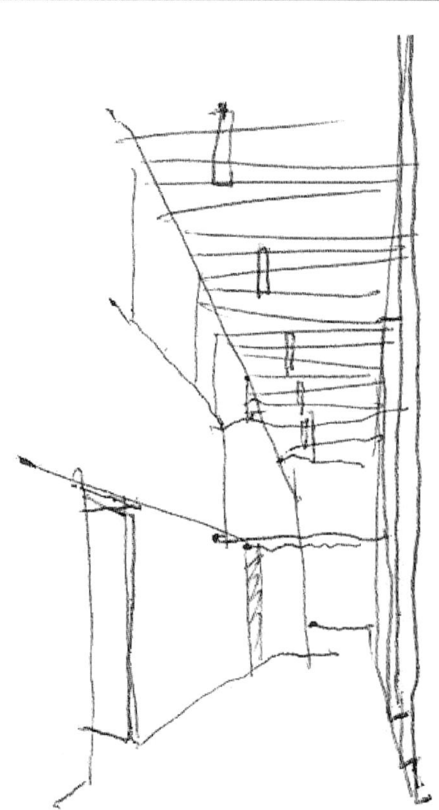

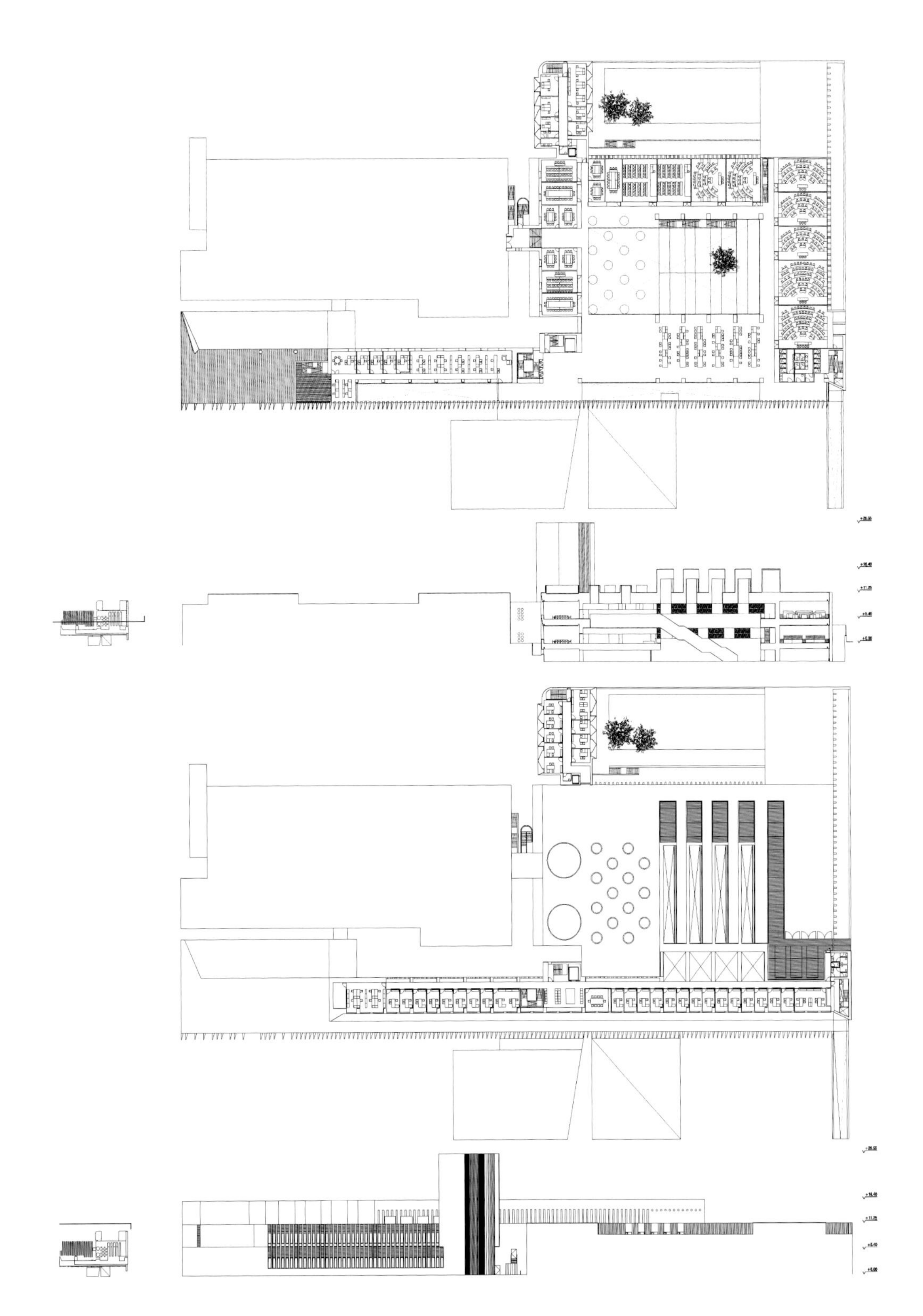

The new building is the definite headquarters of the Faculty of Economy and Business; it has a large provision of classrooms and rooms available for the increase in the teaching possibilities of the University, adapted to its evolution. It is located in the Pamplona campus and it is next to the current Law building, with which it is connected by means of different accesses in the East and South façades.

From the point of view of the shape of the building, we can highlight the desire to respect as much as possible the well-looked after green campus which currently goes down to the river before this building and which it is part, with a special degree of protagonism, of the visual and environmental heritage of the campus.

The construction is attached to the Law building, as mentioned, with a similar size and loyal to its essential alignments and the addition of a new common façade. This is justified by the solution offered for access, and acquires outstanding importance in the result given its exceptional length, and as it it's the one providing the background to the mentioned green campus. It is designed as a façade with a serried structure and rhythmic expression, coherent with its scale. It is made up by a system of vertical concrete elements able to act as a filtering device of space and views.

该新建建筑将成为纳瓦拉大学经济与商业学院的主要教学楼。它靠近现有的法律大楼，在东、南立面上通过不同的出入口设计与法律大楼相互呼应。

在建筑形态方面，建筑设计充分尊重了现有校园的绿色格调，建筑前方绿草如茵的绿地一直延伸至小河边。浓浓的绿意是该校园环境的设计特色。

就像前文提及的，该新建建筑与法律大楼相互呼应，与其规模相当、平面布局类似。建筑外立面被设计成富有节奏感的片层结构。这些结构主要由一系列的垂直混凝土元素打造而成，具有装扮空间和遮挡部分视野的作用。

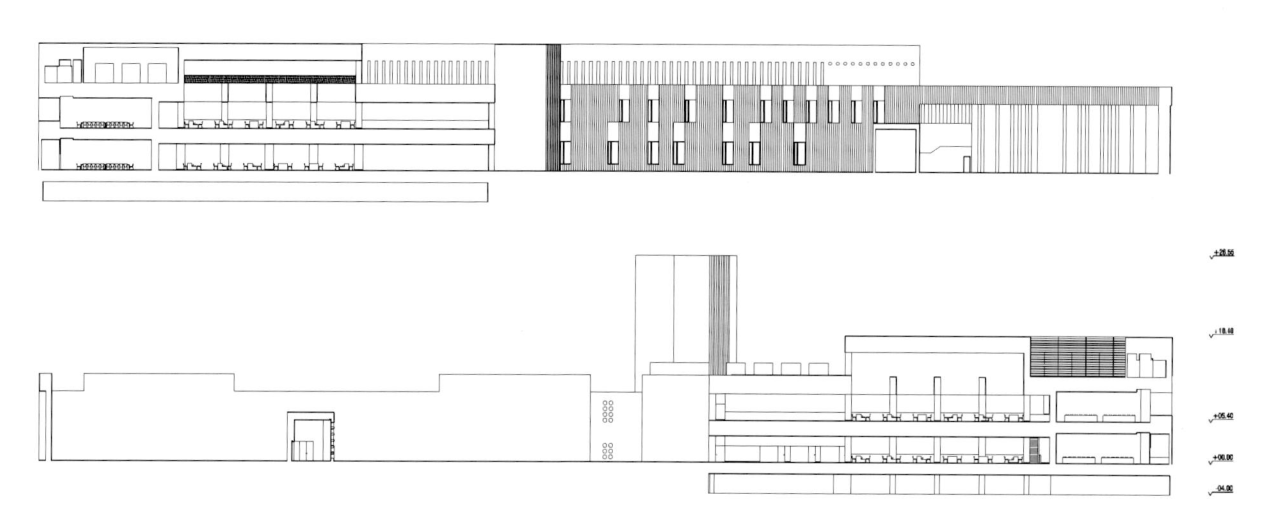
+26.55
+18.48
+05.40
+00.00
-04.00

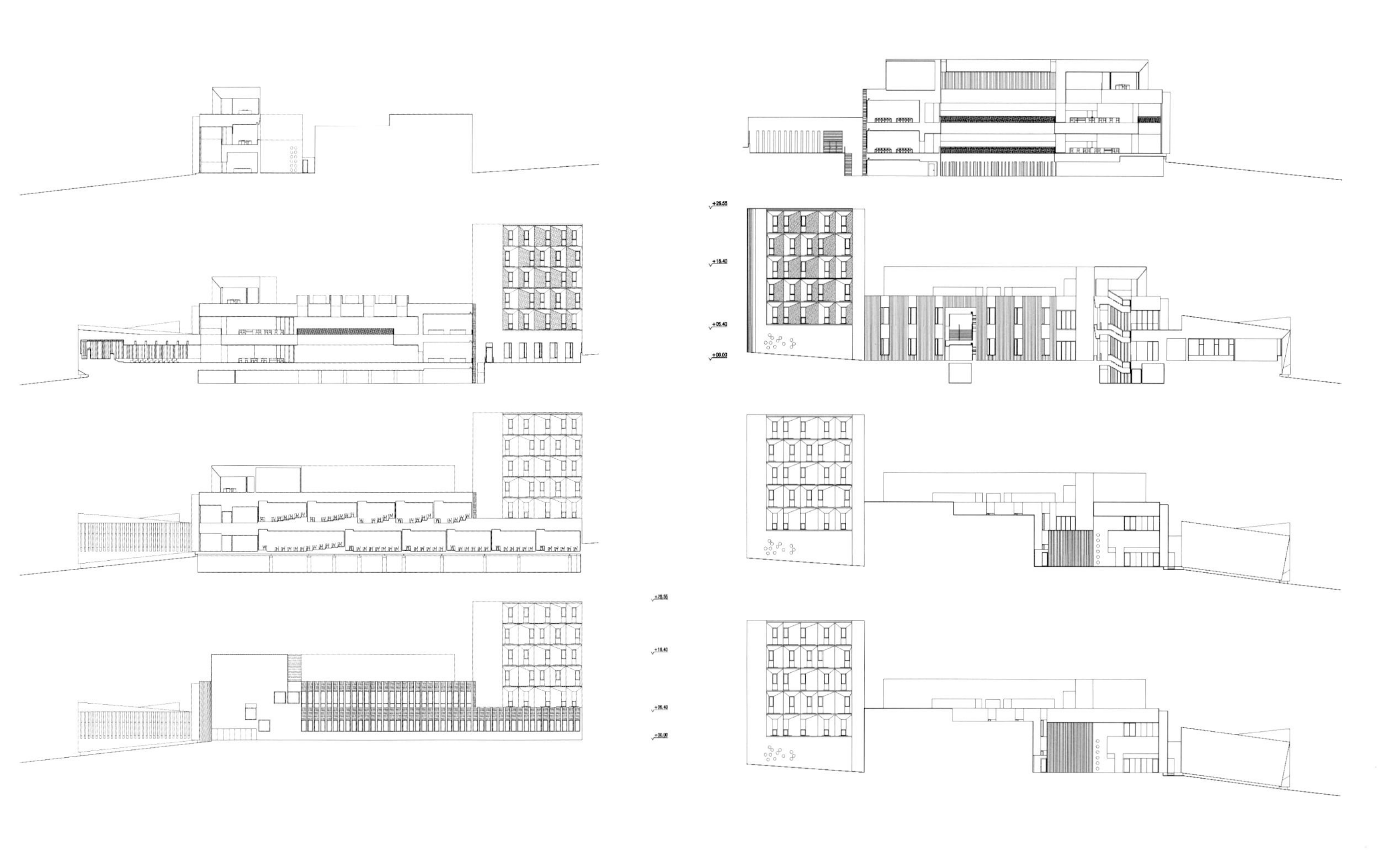

25
ECO
NÓMI
CAS

KC Grande Resort & Spa-Hillside

KC格兰德度假村酒店及水疗中心

Design Company / 设计事务所:
FOS [Foundry of Space]
Project Architect / 设计师:
Makakrai Jay Suthadarat, Rinchai Chaiwarapon, Singha Ounsaku
Location / 地点:
Thailand （泰国）
Area / 面积:
8 000 m^2
Photography / 摄影:
Teerawat Winyarat

视觉亮点

建筑正立面采用了整齐、统一的长方形结构。木质材料和玻璃材料的巧妙搭配，为建筑内部空间提供了良好的观景视野。

Situated at the corner of the inclined road that cuts the site into a triangular shape which is less than 100 meters away from the beachfront, KC Grande Resort & Spa New Extension faces two problematic challenges: how can the building fit in a triangular site with a steady inclination? And how can the guests feel like being part of the beachfront despite the fact that they are far-off?

By splitting the building into two parts in two different levels according to the existing topography, we optimise the internal space between the 5-metre different levels of the front and the back buildings while also strategically placing a 30m-long pool and a 3.5m-high artificial waterfall as a focal point in middle of the space. On the first floor of both the lower and upper buildings, all guestrooms have direct access to the plunge pools of their own, representing the effect of being adjacent to the sea. Another advantage of locating the back building on top of the slope is that most of the guestrooms have clear view towards the sea, not being blocked by the lower building and other buildings in front.

KC 格兰德度假村酒店及水疗中心的扩建项目位于一条倾斜道路的转角处，这条路将地块分隔成一个三角形。地块距离海滩不到 100m。设计师面临两大挑战：建筑物如何在一个倾斜的三角形地带上“扎根”？地块距离海滩有一定的距离，应该如何让游客感觉仿佛置身于海滩环境呢？

设计师根据原有地势将建筑分为高度不同的前、后两个部分，两者高差 5m，充分利用前、后楼之间的内部空间，同时在此设置了 30m 长的游泳池，以及作为空间中央焦点的 3.5m 高的人工瀑布。前部建筑一层的所有房间直接连接游泳池，每个房间都有进入泳池的入口。后部建筑位于斜坡上，因此，大部分房间都拥有眺望大海的良好视野。由于前后建筑有高低落差，因此前部的建筑不会遮挡后部建筑中游客观赏美景的视线。

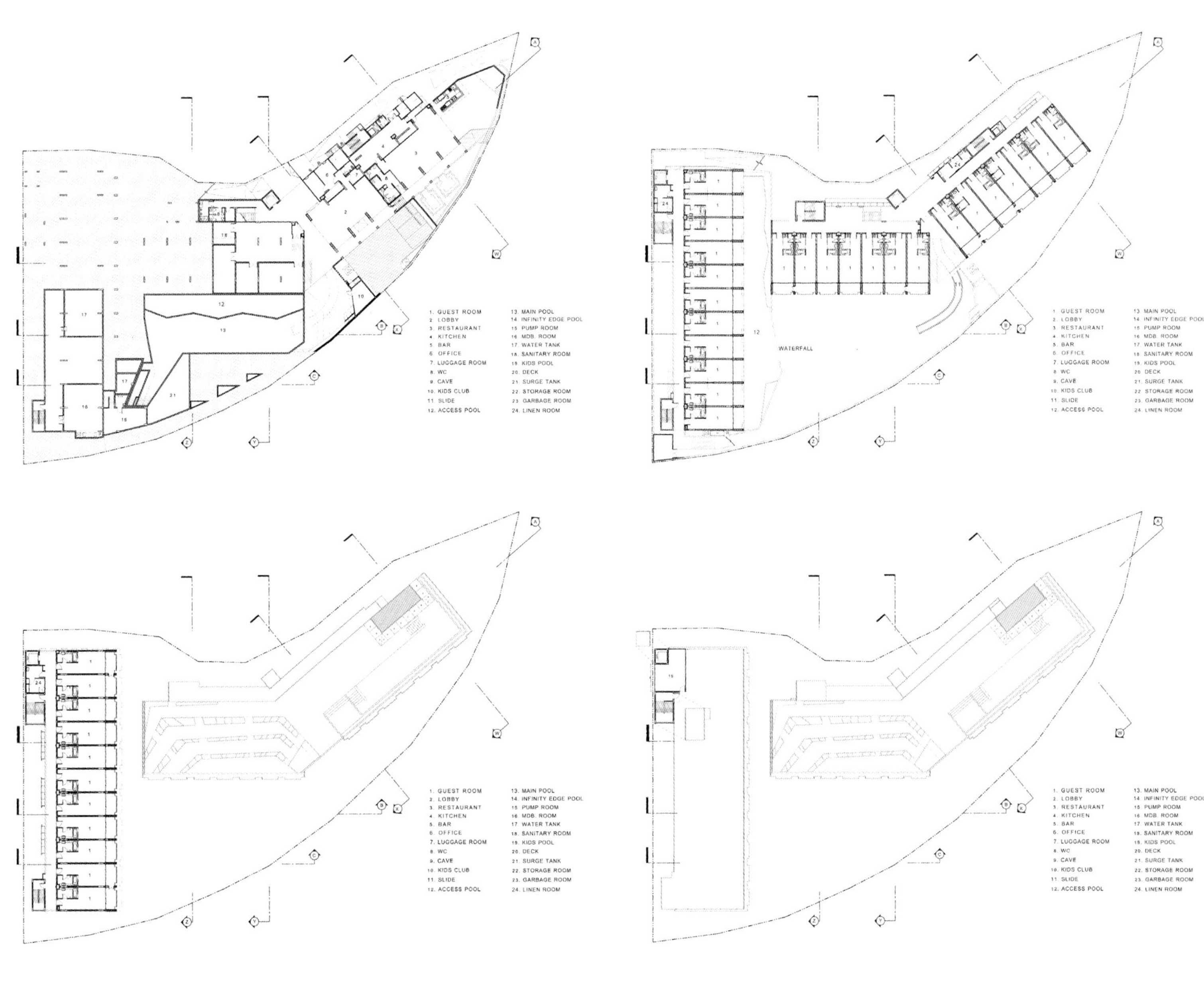
1. GUEST ROOM
2. LOBBY
3. RESTAURANT
4. KITCHEN
5. BAR
6. OFFICE
7. LUGGAGE ROOM
8. WC
9. CAVE
10. KIDS CLUB
11. SLIDE
12. ACCESS POOL
13. MAIN POOL
14. INFINITY EDGE POOL
15. PUMP ROOM
16. MDB. ROOM
17. WATER TANK
18. SANITARY ROOM
19. KIDS POOL
20. DECK
21. SURGE TANK
22. STORAGE ROOM
23. GARBAGE ROOM
24. LINEN ROOM
WATERFALL

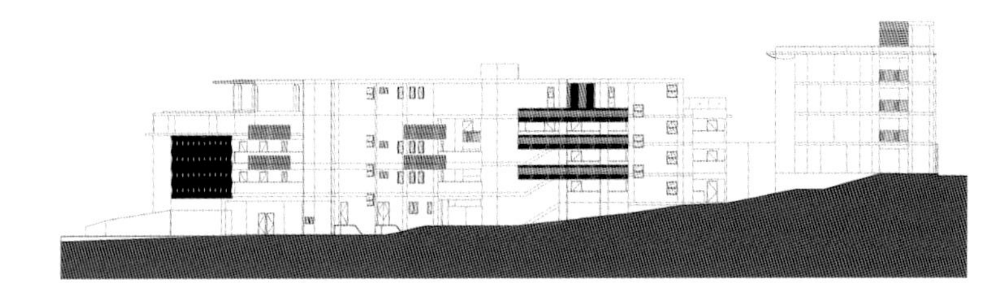
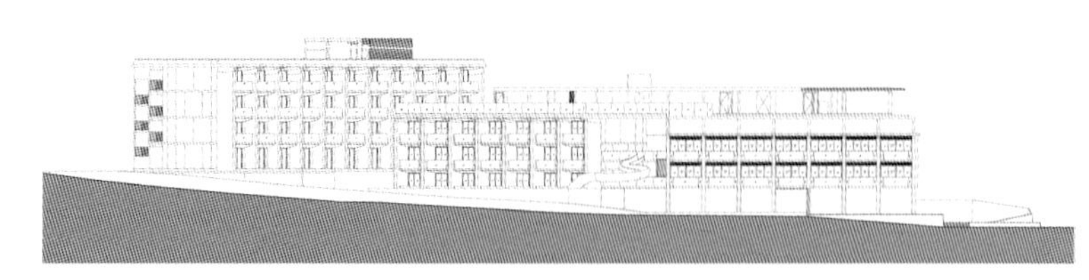
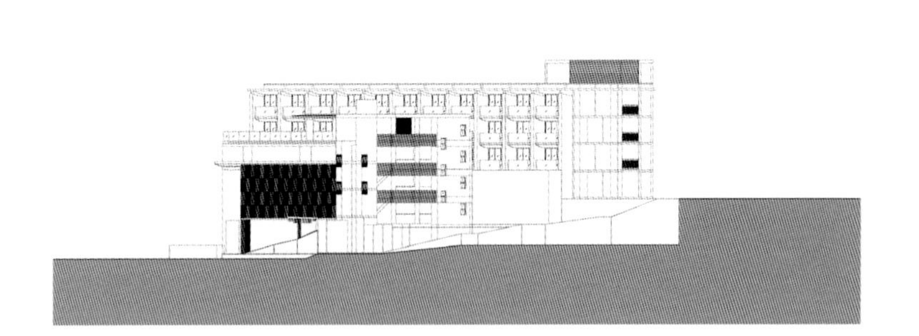
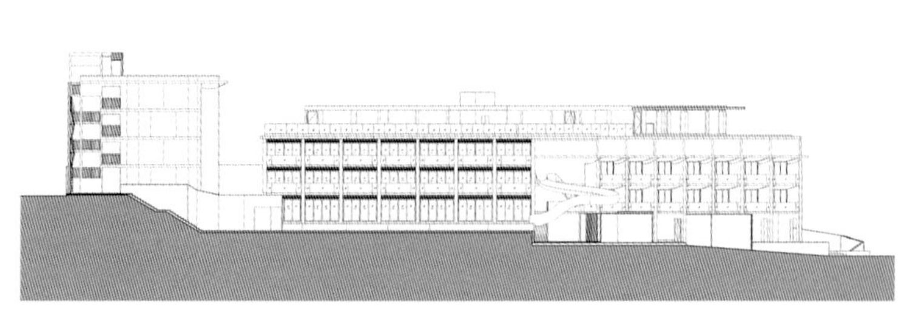

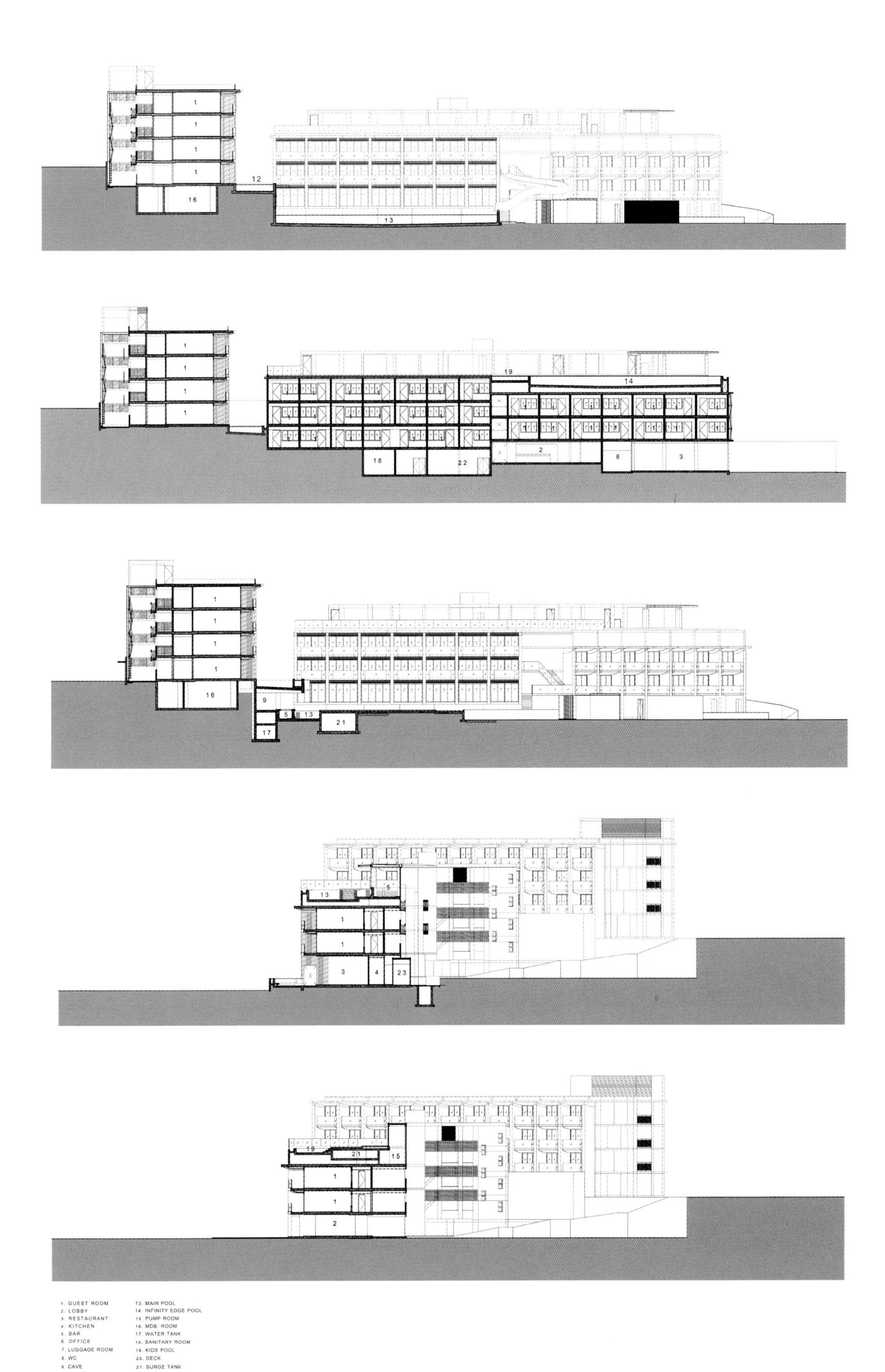

1. GUEST ROOM
2. LOBBY
3. RESTAURANT
4. KITCHEN
5. BAR
6. OFFICE
7. LUGGAGE ROOM
8. WC
9. CAVE
10. KIDS CLUB
11. SLIDE
12. ACCESS POOL
13. MAIN POOL
14. INFINITY EDGE POOL
15. PUMP ROOM
16. MDB. ROOM
17. WATER TANK
18. SANITARY ROOM
19. KIDS POOL
20. DECK
21. SURGE TANK
22. STORAGE ROOM
23. GARBAGE ROOM
24. LINEN ROOM

条形建筑

教育建筑

Binissalem School Complex

巴利阿里学校综合体

Design Company / 设计事务所:
RIPOLLTIZON
Project Architect / 设计师:
Pep Ripoll, Juan Miguel Tizón
Location / 地点:
Spain （西班牙）
Area / 面积:
3166 m^2
Photography / 摄影:
José Hevia

视觉亮点

白色立面干净、简洁，部分彩色的带有开窗设计的外墙点缀其中，使建筑外观充满趣味性，更贴近使用者的心理需求。

30

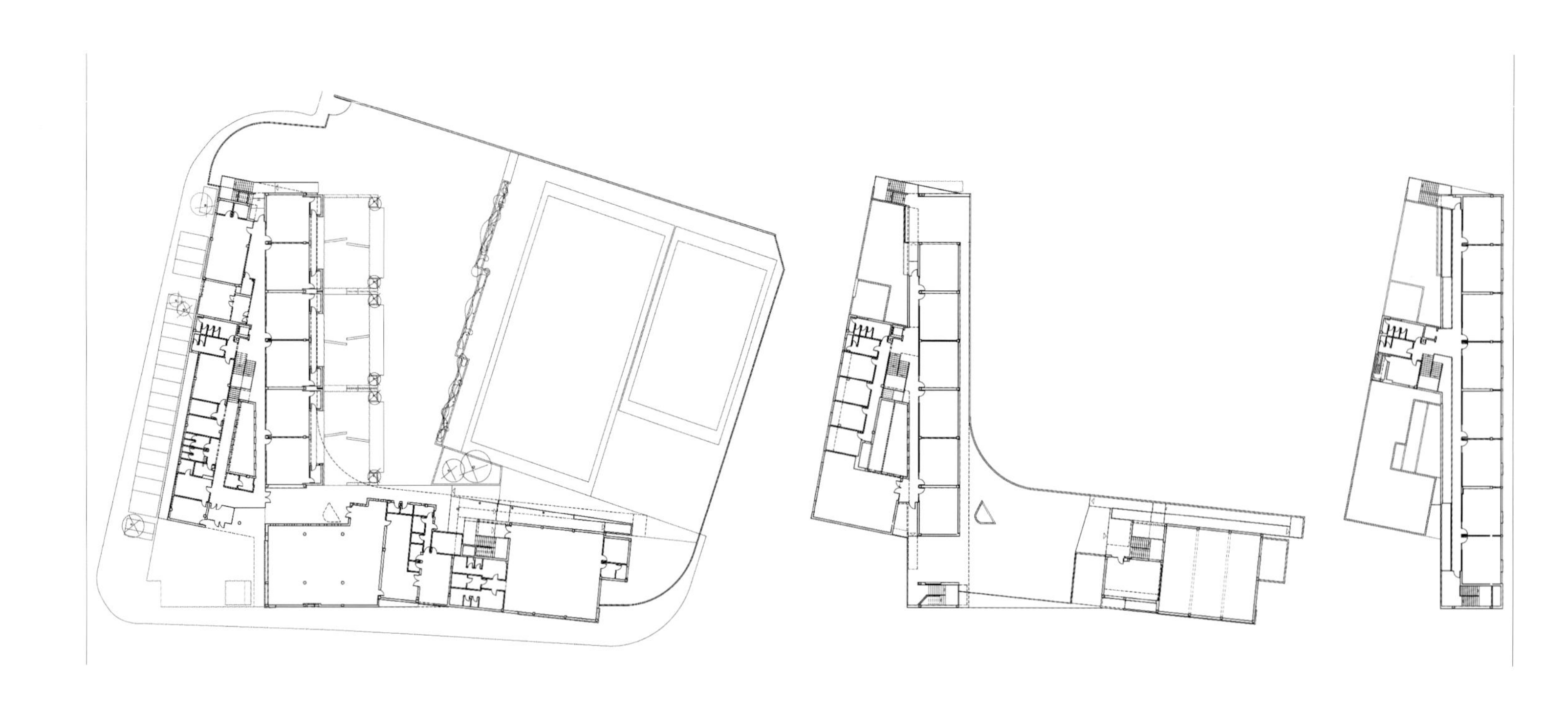

CEIP

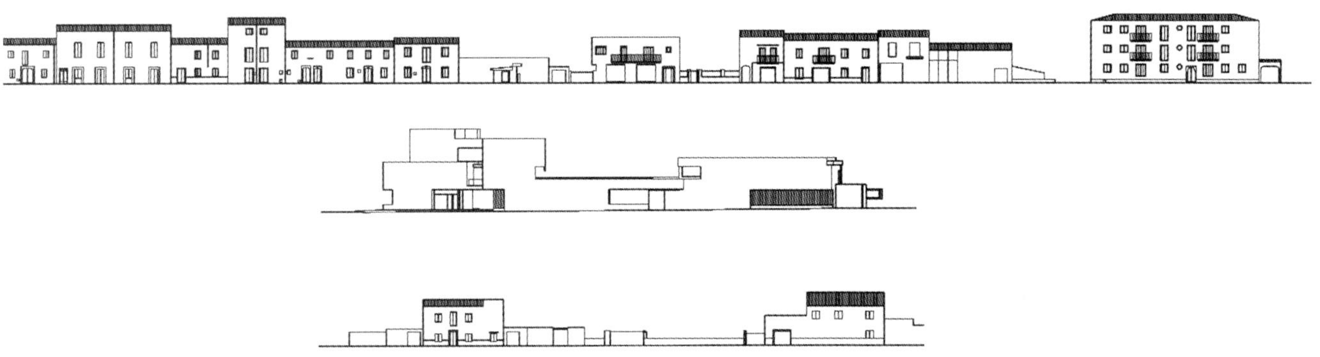

The School Complex (provides primary and secondary school levels) is located in the outskirts of Binnisalem urban fabric. The plot is located along a suburban road named "Camí de Pedaç" on which the urban planning has concentrated a heterogeneous mix of typologies, including diverse row houses, detached blocks and urban facilities.

From the beginning, our intent was to develop the project as a dialogue, on different scales, between the school and its surroundings. On the one hand, the new school building faces the road with a fragmented volume and a broken skyline that enhances perspective effects and scale control in relation to the singularities of the unorganized neighborhood volumes. On the other hand, towards interior of the plot that faces the countryside, the building embraces the sport ground areas creating a façade with bigger scale elements and more compact massing. Moreover, the building areas used only for teaching were clearly separated from those that can be used also for non school events.

A set back on the façade to the road creates the main access space, an open plaza in the building corner, that generates the circulations and arranges the different functions. The functional packages are grouped in different levels with the intention to reduce the building coverage surface and create a plot area where play grounds, sport grounds and future extensions can be located. From this roof plaza is also possible to enjoy the excellent views of Binissalem skyline and its surrounding mountains.

这所学校（包括小学和中学）建筑位于巴利阿里的市郊。地块位于郊区的一条公路边。这里的城市规划集中了不同的结构类型，包括各式各样的联排建筑和独栋建筑。

从一开始，设计师的目标是建造一个能与周边不同尺度空间相呼应的建筑。一方面，新学校临街而建，建筑打破了原有的天际线，增强了透视效果；另一方面，建筑朝向乡村地区，其中包含运动场地，设计师因此设计了一个构件规格更大、体量更为小巧的立面。此外，仅用于教学的建筑领域与安排非学校活动的区域被明显地分割开来。

建筑与公路间有一块空地，这里不仅是一个主要的入口空间，还成为建筑转角的开放广场，形成了交通循环路线，并形成不同的功能区。不同的楼层具有不同的使用功能。设计师还创建了一个游戏场地、体育场地和未来的扩建场地均可涵盖的规划空间。在建筑屋顶广场上也可以欣赏到优美的天际线及其周边山脉的美好风光。

条形建筑

商业建筑

LE 2-22, QUARTIER DES SPECTACLES, MONTRÉAL

蒙特利尔建筑

Design Company / 设计事务所:
ÆDIFICA
Location / 地点:
Canada（加拿大）
Area / 面积:
3809 m^2
Photography / 摄影:
Stéphane Brügger

视觉亮点

富有创意的双层墙面设计和向内收缩的三角形入口使建筑极富标志性。

SUBWAY

CLUB SODA

In addition to its innovative double wall, what strikes you immediately about the building is its recessed, angular entrance. An echo of buildings that stood at this intersection in earlier times, this design frees up the sidewalk for the dense pedestrian traffic this area generates, particularly during festivals. The unique façade is marked by seemingly random openings in the building's wooden envelope. The envelope is protected by a shell of transparent glass, like a second skin; the space created by superimposing these materials will be used for a variety of multimedia installations. The distribution of windows reflects the activities going on inside: the exhibition rooms in the upper levels enjoy controlled lighting, while large openings flood work environments with natural light. Ideas of fluidity and circulation influenced the architecture and create a dialogue between the building and passers-by. In summer, a portion of the envelope retracts at the level of St. Catherine St., altering the building's physical limits. The building was designed as part of "Imagining-Building Montréal 2025". The Angus Development Corporation ensured its cultural vocation by allocating 75% of the space to cultural organizations.

除了富有创意的双层墙面设计，该建筑最引人注目的设计元素在于其嵌入式、棱角分明的入口。该建筑很好地延续了曾经矗立在这个十字路口的老建筑的风格。新建筑为人行道空出了一定的空间，很好地应对了该区域拥挤人流所带来的一些问题，节日期间效果尤为明显。建筑拥有非常别致的立面，其设计的一大亮点在于木质结构上看似随意的窗口设计。透明玻璃幕墙将立面保护起来，是第二层立面。两层立面的中间区域可用来设置丰富多彩的多媒体设施。窗户的分布与空间内的设置相辅相成：建筑上层的展览区拥有可控的光照，而大型的窗口设置使工作区拥有充足的自然光。

流畅性和循环性的理念对建筑设计具有很大影响，使建筑本身与路过的行人进行“对话”。该建筑是“蒙特利尔 2025 年意象建筑”的一部分，建筑事务所将 75% 的空间提供给文化组织。

Pfizer Canada Inc. Siège Social

加拿大辉瑞公司总部办公楼

Design Company / 设计事务所:
Menkès Shooner Dagenais LeTourneux Architectes
Project Architect / 设计师:
Anik Shooner
Location / 地点:
Canada （加拿大）
Area / 面积:
21 160 m²
Photography / 摄影:
Stéphane Groleau

视觉亮点

建筑的外立面选用了金色的金属材料，使建筑获得了丰富的光影效果。

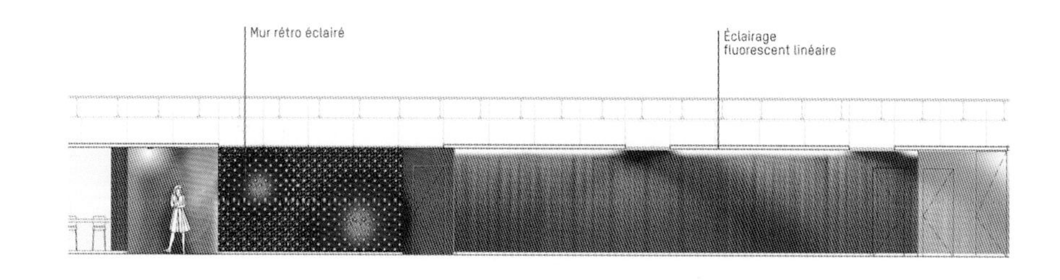

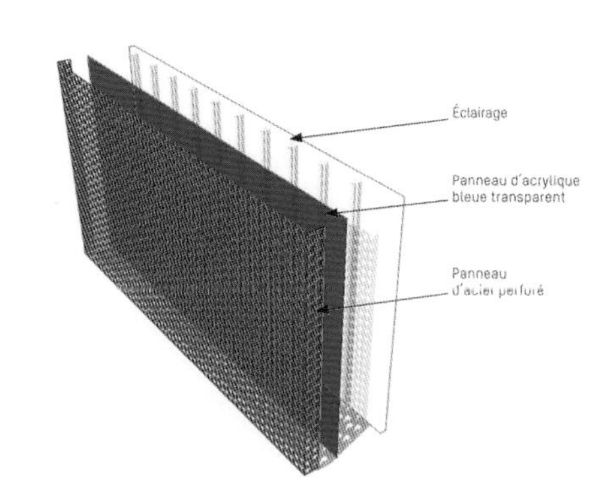

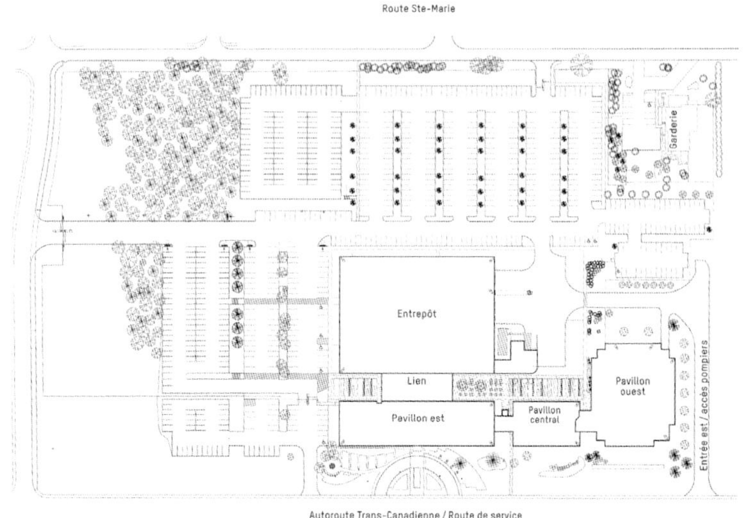

The alveolate structure has been transposed to create an elegant motif displaying a variety of textures that responds to the changing light. The screen openings of the facade reproduce both the scale and the texture of the local brick, which are enriched through a wide range of colors ranging from silver to gold. The blue of the Pfizer logo is also very visible and create a recurring theme; it occurs both on the external facade and in the core of the building. A bond is created between the industrial park, the highway, the company and the building space. Vivid and contemporary, the facades give expression to the unity of the whole. Inside, the redefinition of the site offers brand new, healthy and inspiring workspaces leading the way to a greater economic impact while attracting utmost qualified employees.The construction materials used, such as clear and opaque glass, aluminum and wood, reflect ambient light to create texture. The distribution and arrangement of these primary materials also highlight the multiple functions of various spaces. The entrance way to the main building offers luminous integration to the new façades which act as curtain walls. The design concept is based on the enhancement and the democratic access to natural light, in order to ensure that it illuminates each work station.

设计师通过调换顺序的方式对建筑结构进行重新打造，建造了可随光线变化而变化的丰富多彩的建筑空间。立面为孔状，并采用本地生产的金属材料。银色、金色等丰富的色彩使建筑更具韵味。辉瑞公司的蓝色标志相当醒目，进一步体现了建筑的主题。这种蓝色标志不仅出现在外部立面，也出现在建筑的内部。设计师将建筑本身与工业园、高速路联系起来。建筑富有活力，又不失现代色彩。

建筑还使用了透明或不透明的玻璃、铝合金、木材等材料。在阳光照射下，这些材料会产生不同的光影效果。对这些材料的分布和设置都是为了满足不同空间的使用功能。建筑一侧立面为玻璃幕墙结构，主建筑的入口与立面完美地融合在一起。该项目的一个主要设计目标是让建筑尽可能地接受自然光的照射，以确保每个工作区都能拥有良好的光照条件。

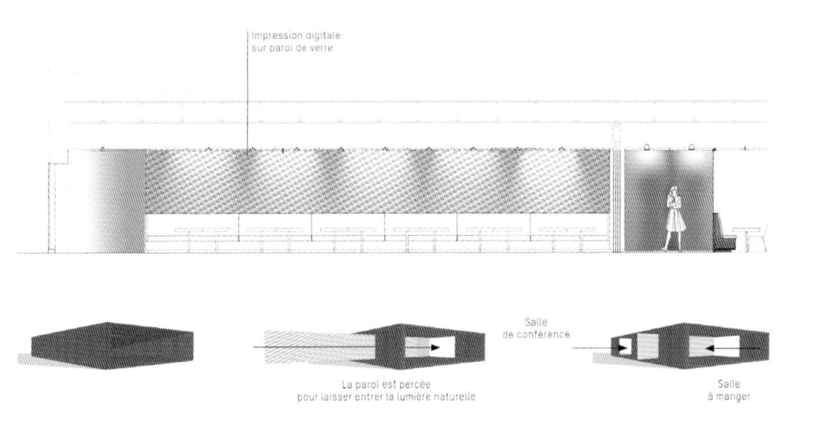
Impression digitale
sur paroi de verre
Salle
de conférence
La paroi est percée
pour laisser entrer la lumière naturelle
Salle
à manger

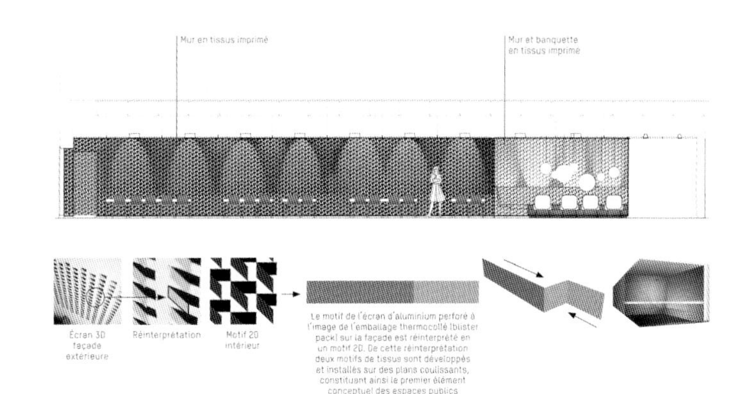
Mur en tissus imprimé
Mur et banquette
en tissus imprimé
Écran 3D
façade
extérieure
Réinterprétation
Motif 2D
intérieur
Le motif de l'écran d'aluminium perforé à
l'image de l'emballage thermocollé (blister
pack) sur la façade est réinterprété en
un motif 2D. De cette réinterprétation
deux motifs de tissus sont développés
et installés sur des plans coulissants,
constituant ainsi le premier élément
conceptuel des espaces publics

Pfizer

条形建筑

居住建筑

Apartment No 1 in Mahallat

马哈拉特一号公寓

Design Company / 设计事务所:
Architecture by Collective Terrain (AbCT)
Project Architec / 设计师:
Ramin Mehdizadeh
Location / 地点:
Iran （伊朗）
Area / 面积:
1 590 m^2
Photography / 摄影:
Omid Khodapanahi

视觉亮点

立面特殊的材料和立面多处的开口设置是建筑最具特色的地方。

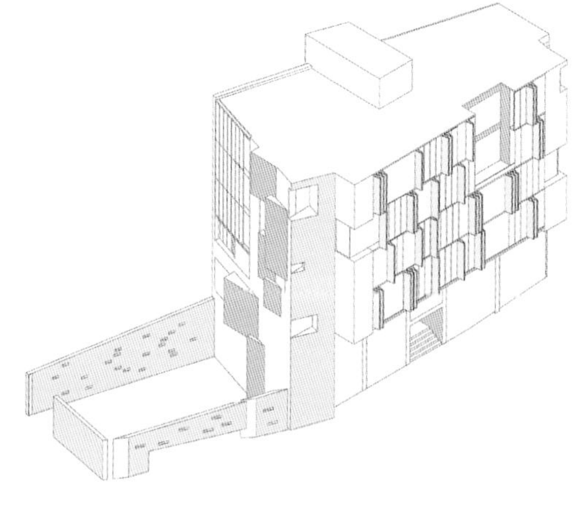

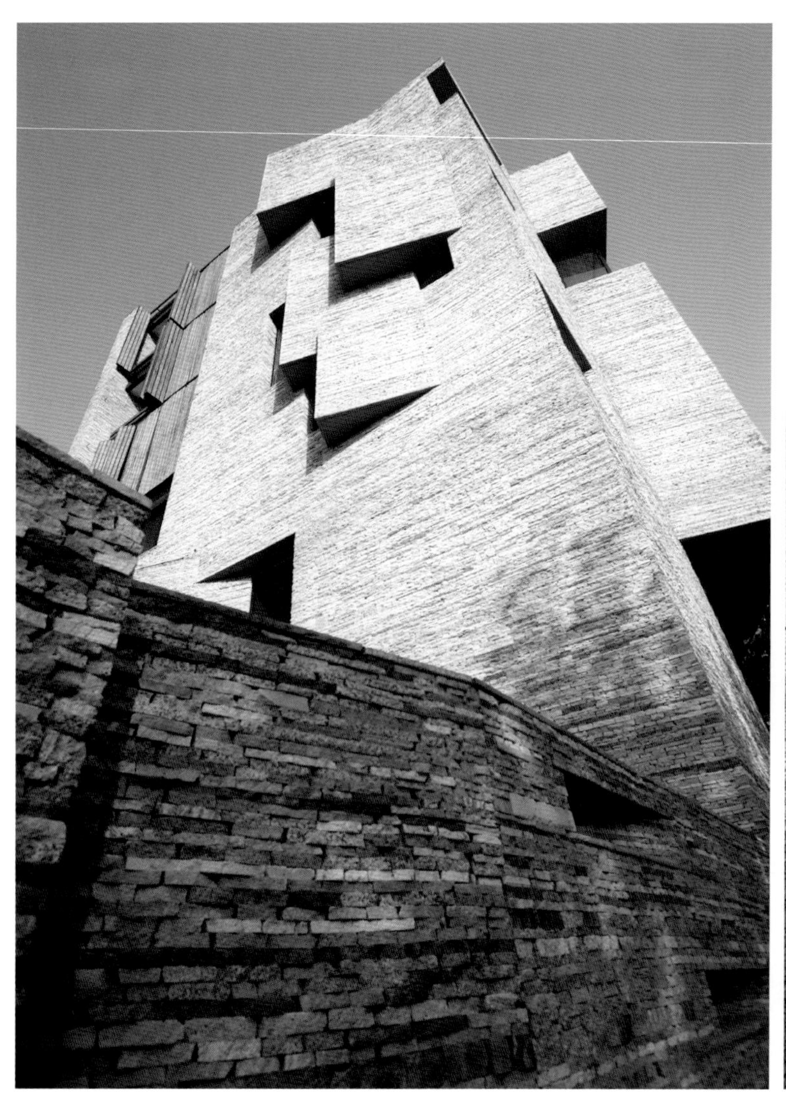

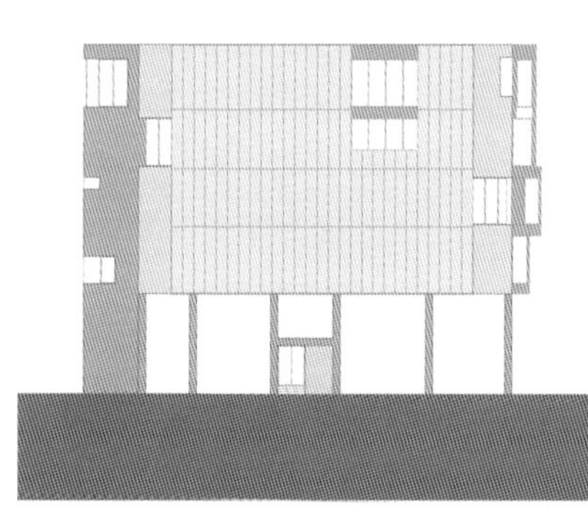

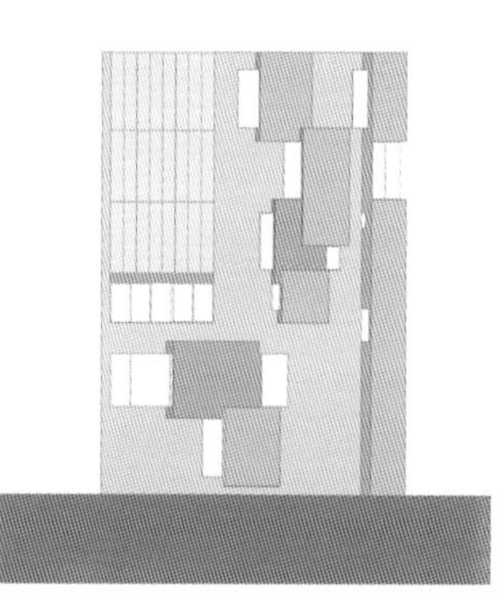

This project is built in Mahallat, an ancient town in the central region of Iran where more than 50% of the local economy is engaged in the business of cutting and treating stones. In this project, leftover stones from the local stone-cutting factories are recycled to use for both exterior and interior walls of the apartment complex. The apartment complex is situated in proximity to the central area of Mahallat, consisting of two retail spaces on the ground level, and eight residential units on the four upper levels.

The building's smooth, austere, abstract prismatic volume is only broken in the areas around the deep-set openings, where triangular additions stick out to protect those apertures. The outer petrous surface is prolonged in the walls inside, using the same rock pieces collected in the local stone-cutting plants, with an immense variety of colors and textures that, because the pieces are so small, is diluted in homogenous surface of the walls.

In the exterior, the larger windows are hidden behind wooden shutters that can be opened during the winter to let the sun shine in and can be closed in the summer to keep it out, still allowing natural ventilation because they are permeable lattices. With such a simple strategy, users regulate lighting and temperature in their homes, and considerable amount of energy saving is achieved.

该项目位于伊朗马哈拉特，这是伊朗中部的一座古老城镇，占当地经济中50% 以上的是切割业和石材加工业。就该项目而言，本地石材切割工厂的废料都被循环利用起来，用来建造公寓的内墙和外墙。该公寓靠近马哈拉特镇中心。建筑一层有两个零售商店，上面四层设置了八个住宅单元。

建筑的棱柱状外观抽象而流畅，设计师只在内凹式的入口处空间设计上使用别的形状，三角式的加建部分向外伸出。建筑立面的石材设计延续到了建筑内部空间中，建筑内部的墙也使用了当地石材切割工厂生产过程中剩下的材料。设计师设置了多种色彩和质地的墙壁，并保持设计的统一性和协调性。

大型的窗户掩藏在木质百叶窗之内，冬天可以打开百叶窗让阳光照入室内，而在夏天又可关上窗户，将强烈的阳光阻隔在外。使用者可以自行调节室内的光照和温度，这节约了大量能源。

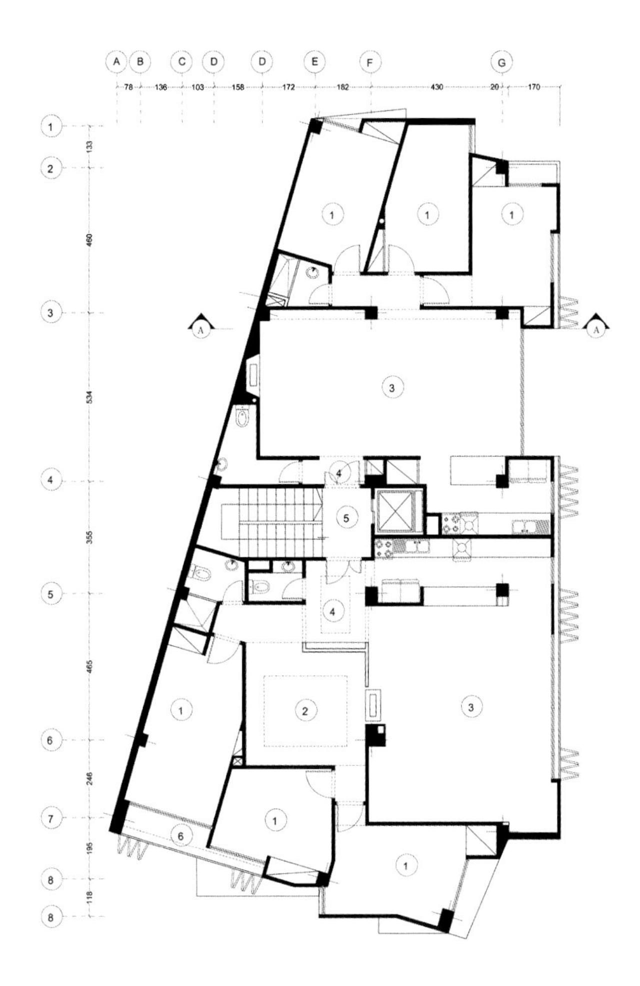

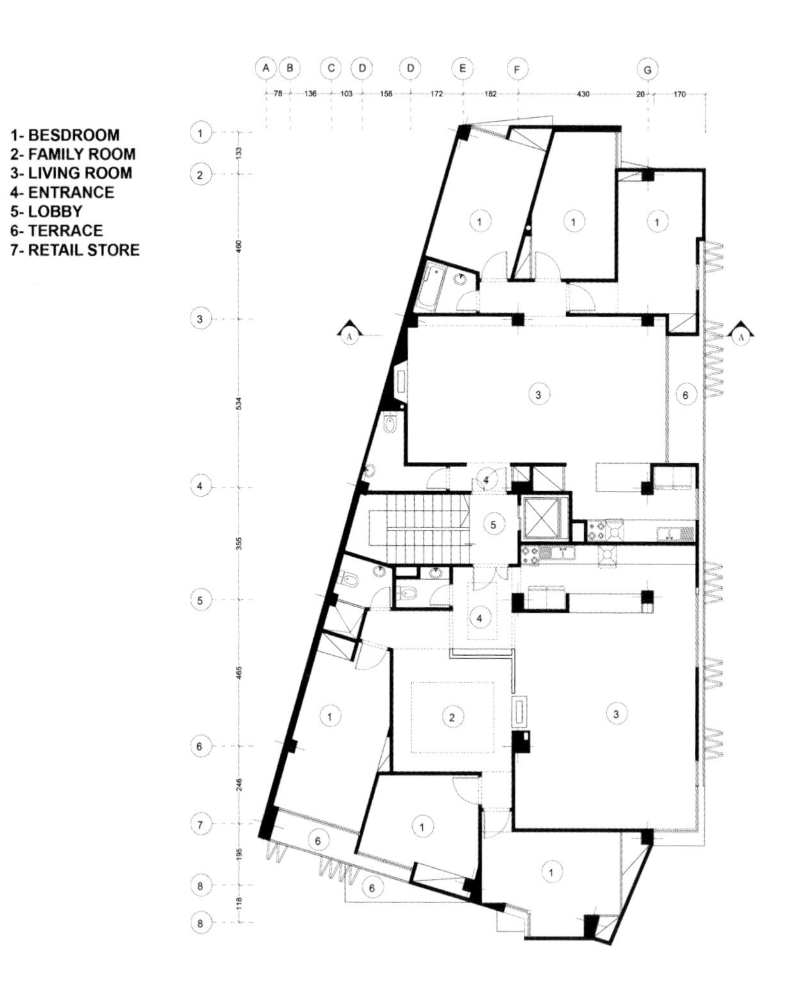
1- BESDROOM
2- FAMILY ROOM
3- LIVING ROOM
4- ENTRANCE
5- LOBBY
6- TERRACE
7- RETAIL STORE

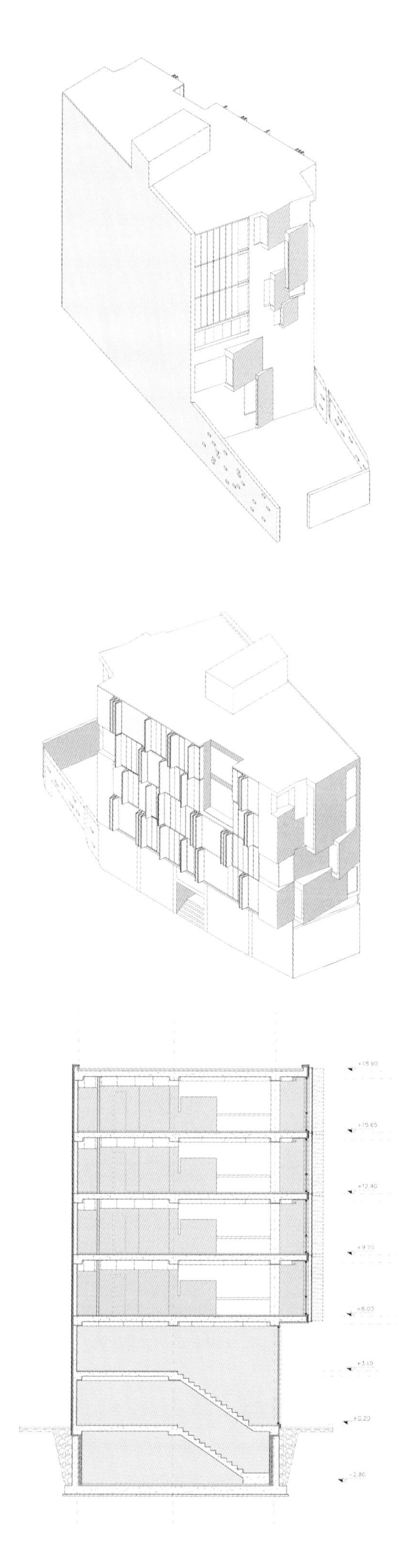
+18.90
+15.65
+12.40
+9.25
+6.00
+3.10
+0.20
-2.80

条形建筑

教育建筑

Giraffe Childcare Center

长颈鹿儿童看护中心

Design Company / 设计事务所:
Hondelatte Laporte Architectes
Project Architec / 设计师:
Virginie Davo (manager), Charlotte Fagart
Location / 地点:
France（法国）
Area / 面积:
1450 m^2
Photography / 摄影:
Philippe Ruault

视觉亮点

动物结构恰如其分地出现在建筑体块中，为建筑增添了趣味性。

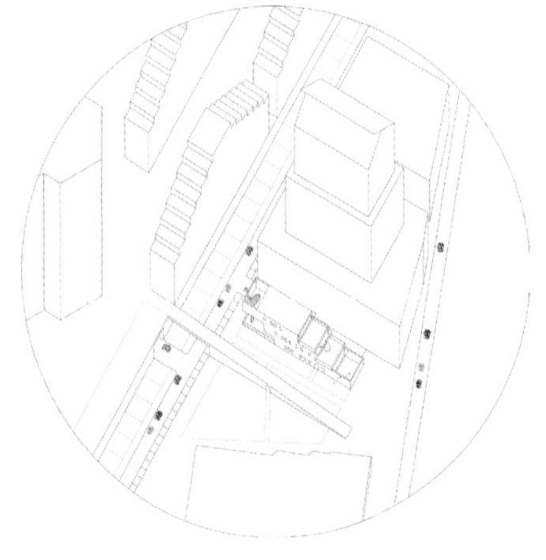

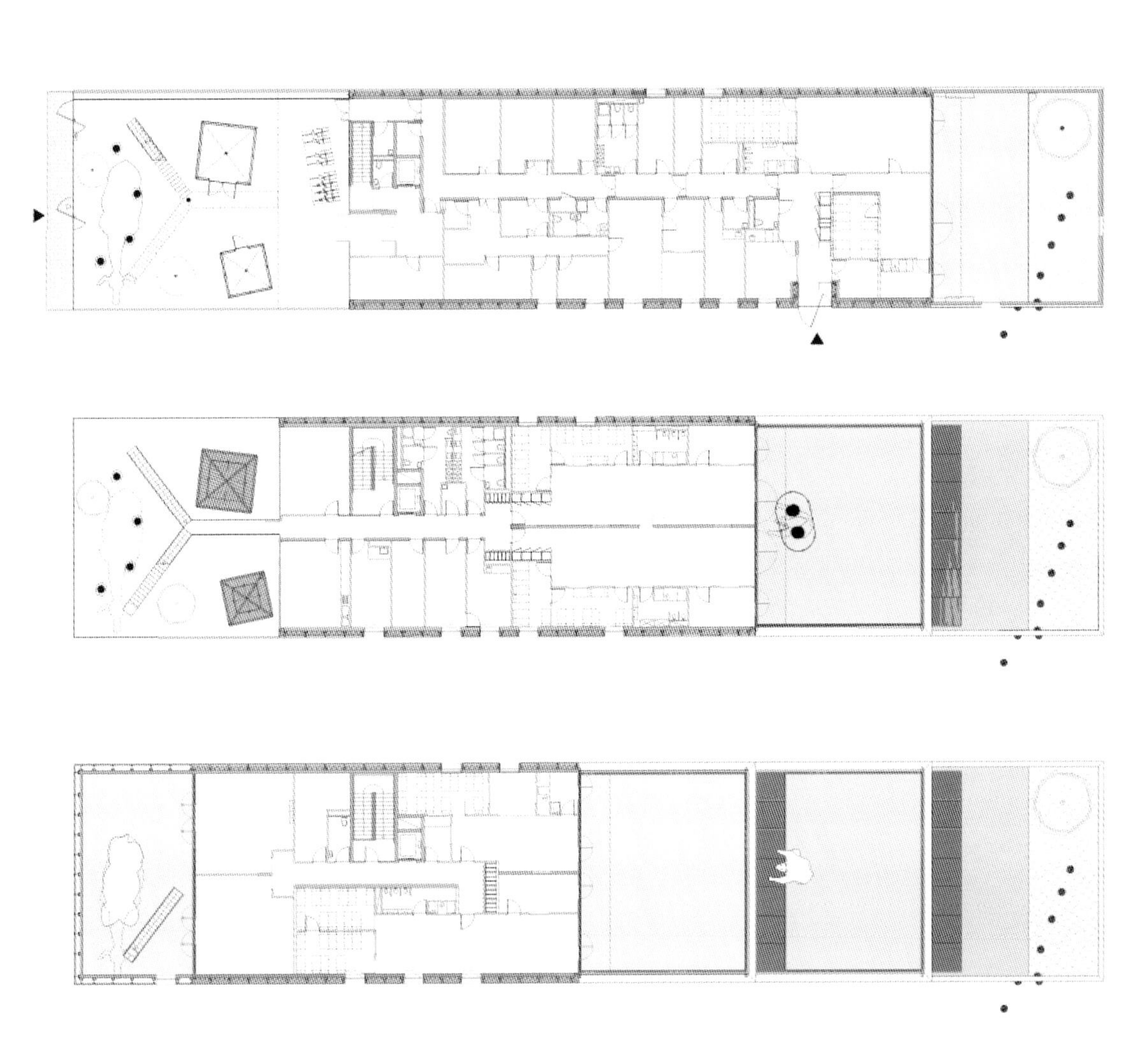

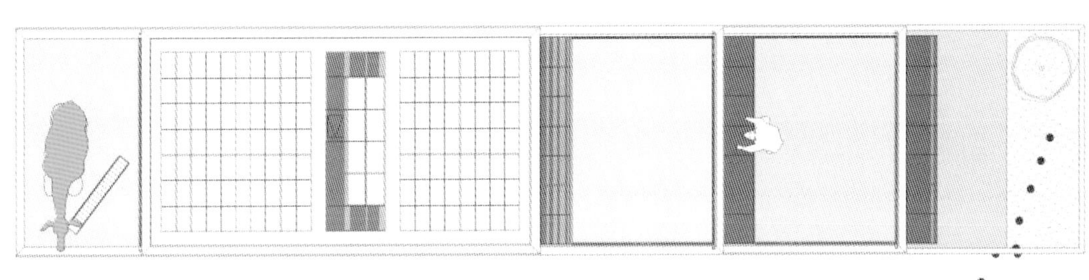

The Giraffe childcare center is located in the C1 block. The program houses a 60 bed childcare center and 20 bed day nursery. The building has been awarded the green "zéro Energie Effinergie" label. The facades of the building are made out of white corrugated iron that provides a minimal background to the wild animal sculptures. The idea is to animate the urban landscape by using a child's imagination. The wild animals appropriate the space; a giraffe appears to be peacefully eating the leaves of the trees from the neighbouring park, a polar bear tries to clamber up the steps, while a family of ladybirds climbs the façade in an attempt to reach the interior patio.

Architecture turns into storytelling. The building changes its identity and becomes a landscape in its own right, a metaphor for the urban jungle. The animals and the trees link the building to nature and motion. We walk through its legs to enter the building. Through their affable form, the lively animal sculptures invite us to live our dreams. These playful and dreamlike sculptures introduce a little bit of fantasy into the routine life of the town in order to inspire our lives with a bit of poetry

长颈鹿儿童看护中心，位于巴黎郊区的 C1 地块上。该建筑包含 60 个儿童看护床位和 20 个婴儿床位。建筑为绿色建筑。建筑外墙的白色波纹板是野生动物雕塑的最佳背景。设计师将公共空间营造出儿童世界的感觉，创造了一些看似荒谬却充满趣味和吸引眼球的元素。一只长颈鹿穿过建筑，北极熊试图从南面爬上屋顶，瓢虫试图从外墙翻进院子里。

建筑好像传达了一个故事。建筑改变了身份，变成都市中的一处景观。穿插在建筑中的动物和树木赋予建筑自然气息和趣味性。人们经过长颈鹿的腿部进入建筑中。这些可爱的动物雕塑使人们产生了有趣、好玩的想象，让人们感受到生活的诗意和美好。

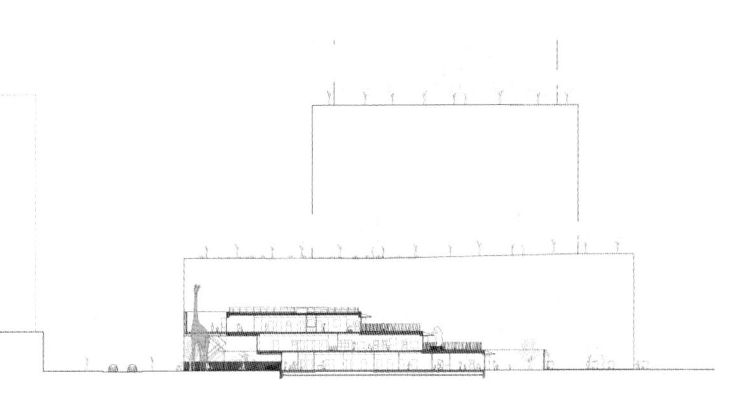

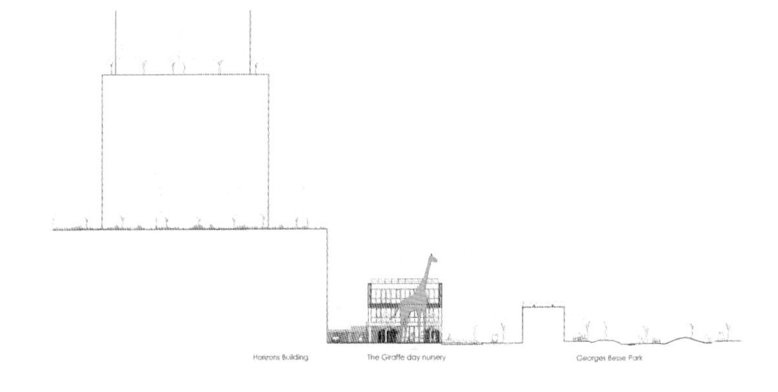
Horizons Building
The Giraffe day nursery
Georges Besse Park

Podcetrtek Traffic Circle

博德森特克交通岛

Design Company / 设计事务所:
ENOTA
Project Architec / 设计师:
Dean Lah, Milan Tomac, Alja Černe, Tjaž Bauer
Location / 地点:
Slovenia（斯洛文尼亚）
Area / 面积:
380 m^2
Photography / 摄影:
Miran Kambič

视觉亮点

设计师将多个混凝土条形结构呈不规则布局。这些条形结构就像是自然生长在地面以下并有着深深的“根系”，它们使地面不停地“膨胀”并挤出裂缝。这种特殊的外观设计使其成为令人难忘的交通标志。

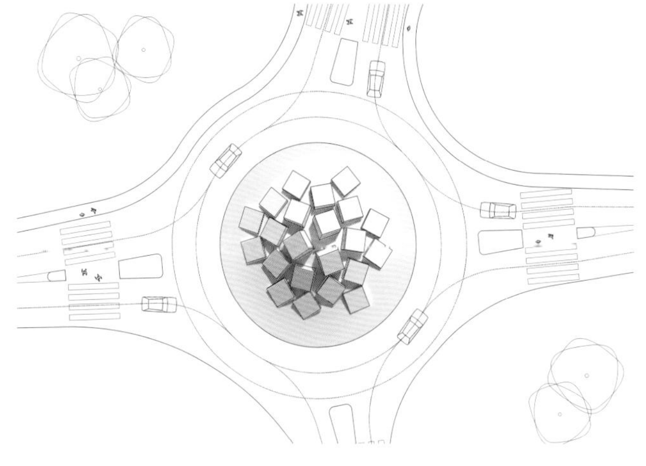

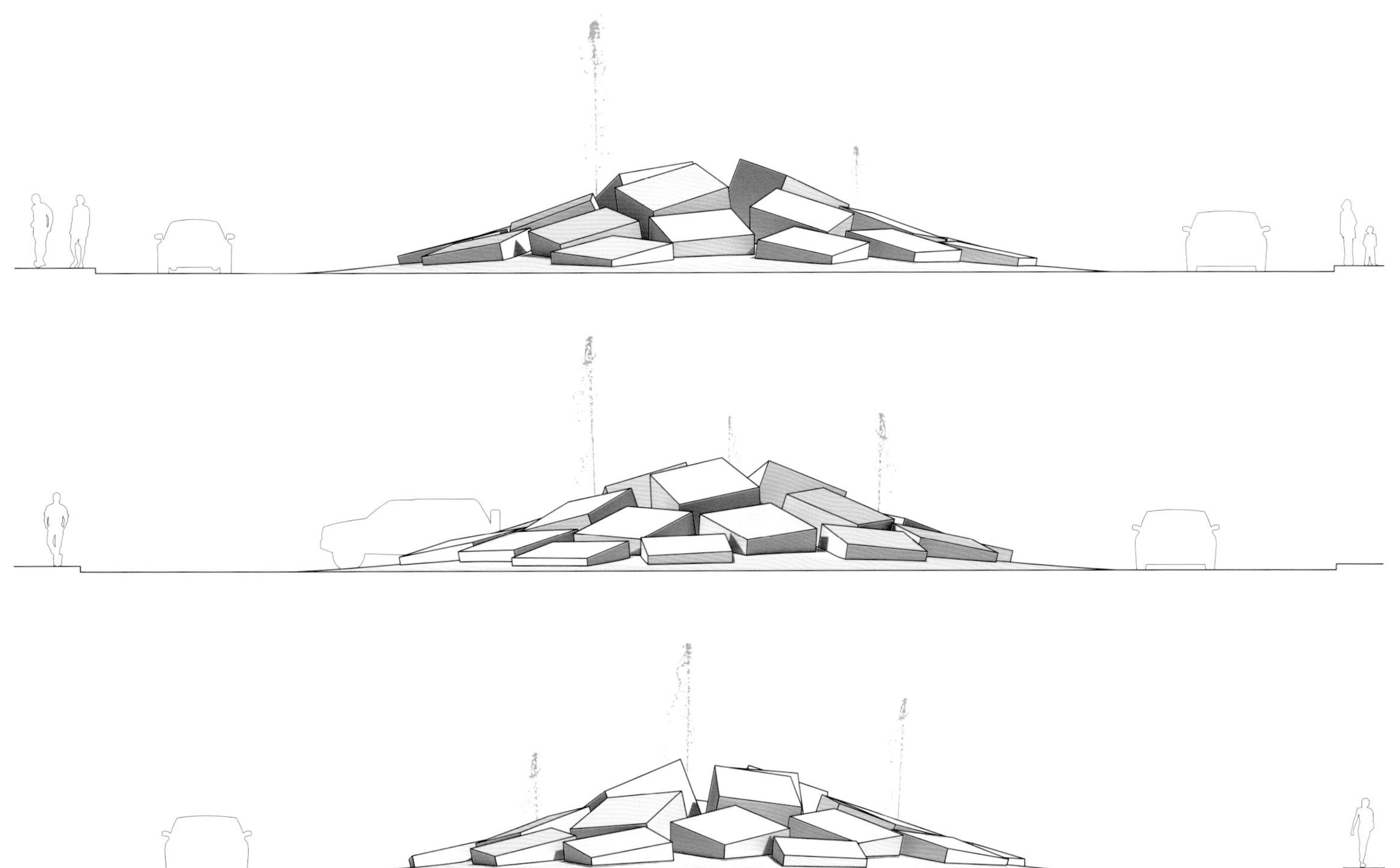

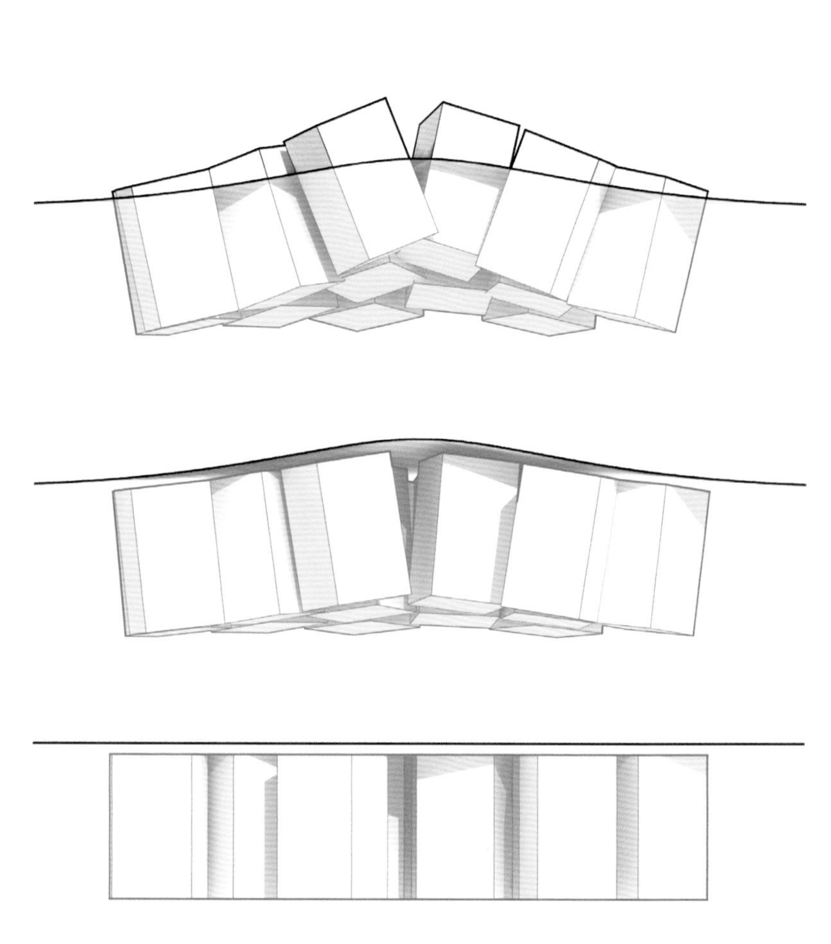

The Podčetrtek Traffic Circle is built on a regional road located between the municipal sports hall with open-air sports grounds on one side and a spa centre with numerous swimming pools and hotels on the other side. It is primarily intended to slow down the traffic in this consequently very busy area as the main accesses to both complexes also connect to the traffic circle. The design of the roundabout's central island thus references the appearance of both facilities and marks the entrance points to the destinations of the visitors to either of the programme centers.

The large, dark concrete blocks allude to the design the monolithic volume of the sports hall. The play of light on the irregular arrangement of the elements forms a composition of surfaces, which corresponds to the expression of the hall's folded volume. The layout of the inner part of the roundabout as a whole suggests a tectonic shift somewhere beneath the Earth's surface having caused the road surface to bloat and belched out the massive blocks. In combination with the water, which sporadically rises to the surface between the clefts, it is somewhat reminiscent of geyser-strewn basalt strata, its appearance thus also evoking the spa complex.

这处交通岛建在当地的一条公路上，其一侧为设有户外运动场的城市体育馆，另外一侧为设有众多游泳池和宾馆的水疗中心。建造交通岛的初衷就是要缓解这里的交通压力，同时保证通向上述两处设施的交通畅通。交通岛的外观设计要和城市体育馆、水疗中心有所呼应。交通岛清晰明确的指向设计会使游客快速、方便地到达目的地。

巨型的黑色混凝土结构会使人们联想到体育馆庞大的外观。交通岛的各个条形结构呈不规则布局。交通岛的这种特殊的外观设计与体育馆的折叠式构造有所呼应。从整体上看，交通岛的结构布局展示了地下的一种构造转变方式：地面向外膨胀，形成一个个的巨型结构。交通岛的设计中加入了水的应用，水通过构造间的裂缝喷洒出来，就像玄武岩地层形成的喷泉一般。这样，交通岛的设计也与附近的水疗设施建立了联系。

条形建筑

商业建筑

Aura Shopping & Entertainment Center

奥拉购物娱乐中心

Design Company / 设计事务所:
Yazgan Design Architecture

Location / 地点:
Russia（俄罗斯）

Area / 面积:
170 000 m^2

Photography / 摄影:
Yunus Özkazanç

视觉亮点

设计师将多个长方形拼凑、组合并应用在建筑外立面设计上，这种外观设计简洁大气。特殊材料和醒目的橙色搭配，使建筑充满了动感。灯光设计不仅满足了照明需求，更使建筑在夜幕下光彩夺目。

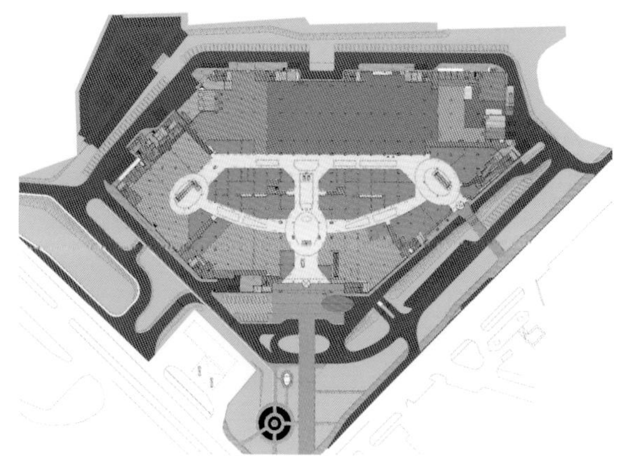

О'КЕЙ
Медиа Маркт
О'КЕЙ

О'КЕЙ
Медиа Маркт
О'КЕЙ
гипермаркет
Снежная Королева
ZARA
Спортмастер
АУРА
CITY
FINN FLARE
TOP SECRET

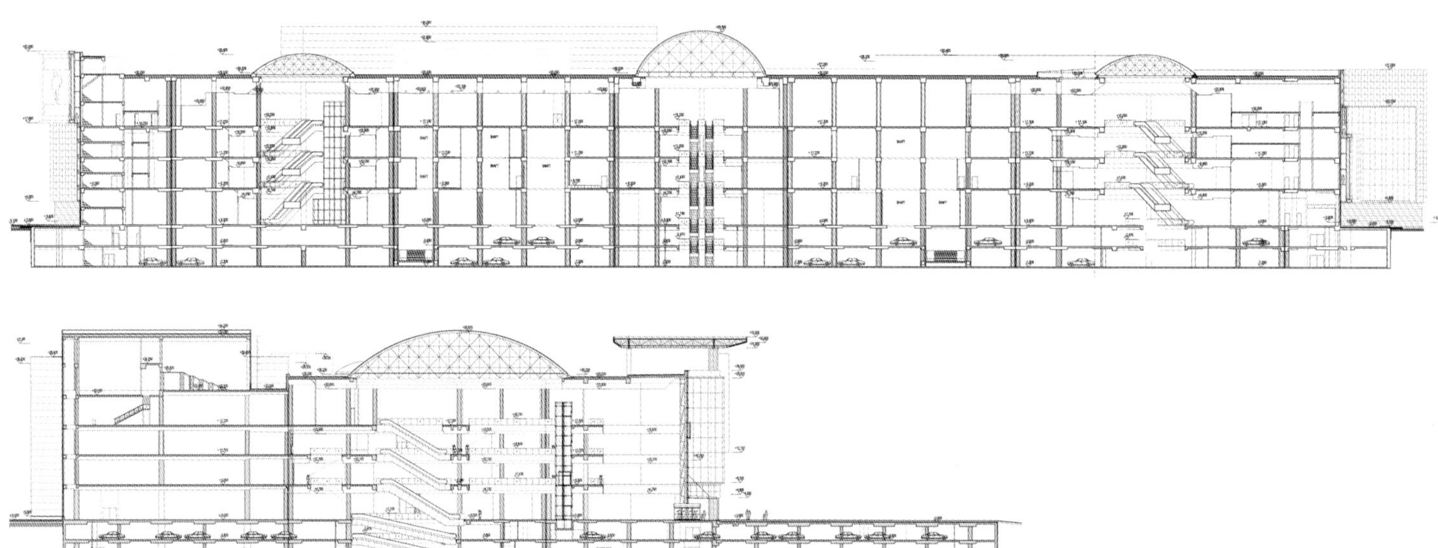

Aura is a 170.000 sqm shopping and entertainment center, which is located in Novosibirsk, Russia. The four storey building has 170 shops with large scaled shopping and entertainment units, such as, a hypermarket, a technology shop, an entertainment center and a cinema, making it the largest shopping center of the city. The building façade was made of enamel painted glass, pattern-imprinted metal panels and light boxes. The glass elements, which are printed with different tones of orange in color, are placed to the façade in angular directions, enhancing the three dimensional characteristic of the facade. Metal panels are imprinted with a particular orange colored flower pattern. The interior columns are covered with the same orange colored flower shaped pattern as of the exterior façade, and helps in developing a connection of the interior design with the exterior architecture. The glass façade and advertisement boxes are lit with LED material, making the building a light sculpture at night.

The building has three main atriums and two secondary atriums that get daylight from five skylights at the top. The skylights are made up of steel elements. The steel sections of three semi sphere shaped skylights, which cover main atriums, are particularly designed according to the shapes of spans, rather than being selected among standardized ones.

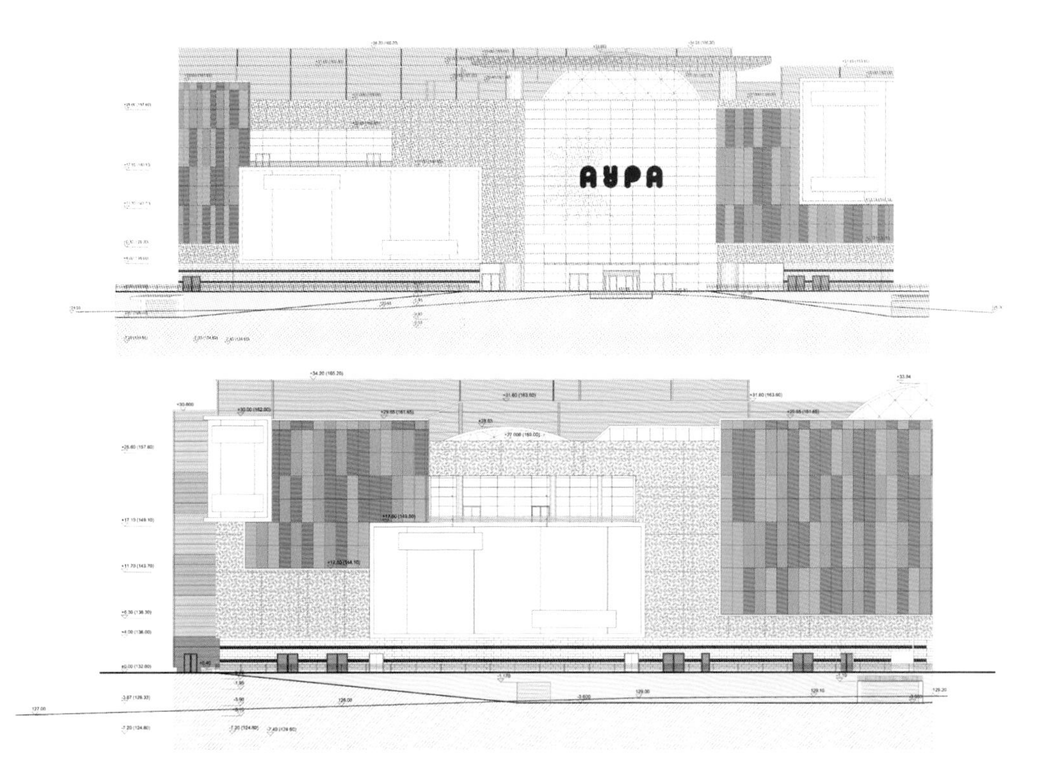

奥拉购物娱乐中心占地170000m^2，位于俄罗斯的新西伯利亚市。这座四层建筑中开设了170多家店铺，其中有很多大型的购物和休闲娱乐场所，这里有一家大型超市、一家娱乐中心、一家电影院，等等。这是这座城市中规模最大的购物中心。建筑立面使用的材料为彩色玻璃、印花金属板材、灯箱，等等。大量的橙色玻璃元素应用在建筑外立面上，提升了建筑外立面的立体感。金属板材部分装饰了风格独特的橙色花朵图案，内部柱廊部分也饰以相同色调的花朵图案，这样，建筑的内外设计就构建起了一种紧密的关联。玻璃立面和广告用的灯箱均使用LED材料进行打造。夜幕降临，整座建筑就像是一座闪闪发光的雕塑，散发着无穷魅力。

该建筑中设有三个主厅和两个次厅，均通过建筑顶部的天窗实现采光。天窗使用钢制元素打造而成。三个半圆形天窗的钢结构呼应着下方的三个主厅。在设计上，钢结构也是依照空间跨度特别定制的。

ZARA
PULL&BEAR

Plein Soleil

巴黎公寓

Design Company / 设计事务所:
rh+ architecture
Location / 地点:
France（法国）
Area / 面积:
2760 m²
Photography / 摄影:
Luc Boegly

视觉亮点

建筑立面上设置了大小不一的阳台，创造了一个整齐、简洁的建筑形态。

SUPERETTE
DU RIQUET

SUPERETTE
DU RIQUET

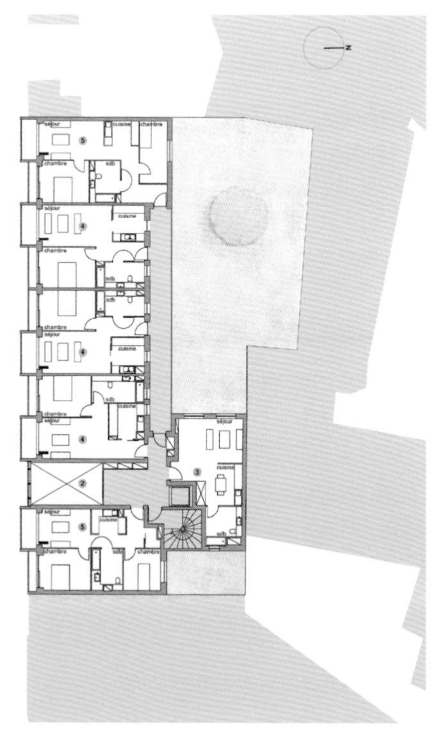

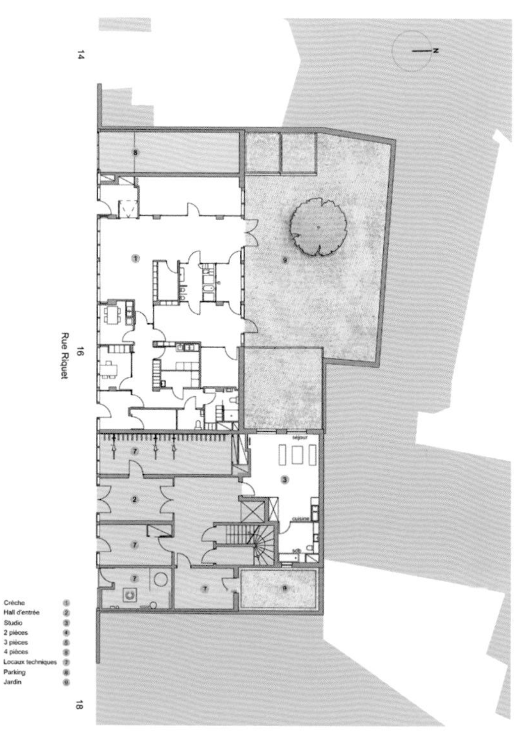
14
16
Rue Riquet
18
Crèche 1
Hall d'entrée 2
Studio 3
2 pièces 4
3 pièces 5
4 pièces 6
Locaux techniques 7
Parking 8
Jardin 9

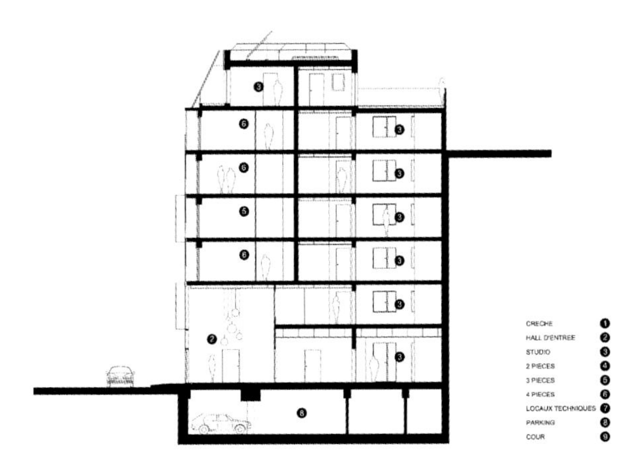

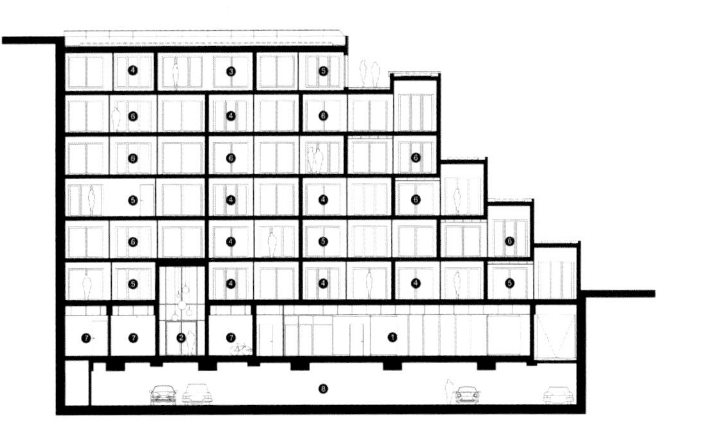

CRECHE 1
HALL D'ENTREE 2
STUDIO 3
2 PIECES 4
3 PIECES 5
4 PIECES 6
LOCAUX TECHNIQUES 7
PARKING 8

The situation of the plot at number 16 rue Riquet is exceptional: largely visible from the corner of Avenue de Flandre, it is very close to the Bassin de la Villette .

The south façade is organized on a principle of loggia. The thermal limit is located at the level of the 30 % opaque and double-glazed very efficient inner façade. The exterior sliding window pane is a simple, slightly printed glazing for the bedrooms and transparent for the living-rooms.

As far as the ground plan and spatial organization are concerned, the qualities of the project are obvious: all flats are through and the bathrooms get daylight. What's more, each flat opens widely on to the south side to capture the most of the sun.

This « thick » façade consisting in loggias running outside along the living-rooms and the bedrooms provide a nice patio area. The depth of these loggias allows tables and chairs to enjoy the sun. As extensions of the living-rooms some of the loggias have clear glass bays on two levels. This extra space can be opened or closed depending on the sunlight. It is both a balcony and a winter garden.

该建筑位于法国巴黎图卢兹大街 16 号，从弗兰德林荫大道的拐角处就可清楚地看到这座建筑。该地块非常靠近维莱特低地广场。

该公寓建筑的南部立面类似于凉廊的设计。通过不透明结构和双层立面，建筑很好地实现了对热能的控制。外部滑动门窗设计非常简单，卧室为半通透式的空间设计，起居室则采用了通透的玻璃结构。

在平面设计和空间组织方面，项目的设计意图非常明显：所有的浴室都能被阳光直接照射到，所有的公寓单元均在南侧设置了大型的窗户，以拥有充足的光照。

沿起居室、卧室设置的“厚厚”的凉廊式立面设计，打造出了优美的阳台区域。作为起居室的延长部分，凉廊拥有两个层次的通透玻璃设计。其中一部分可以依据阳光照射情况打开或者关闭。这里既是阳台，也是冬日花园。

1.2
1

条形建筑

Salmtal Secondary School Canteen

萨尔姆塔尔中学餐厅

Design Company / 设计事务所:
SpreierTrenner Architekten
Location / 地点:
Germany（德国）
Size / 面积:
gross building 552 m^2, schoolyard 1 122 m^2
Photography / 摄影:
Guido Erbring Architekturfotografie

视觉亮点

建筑立面所采用的亮红色的方格式开孔设计，赋予建筑极大的趣味性和可识别性。

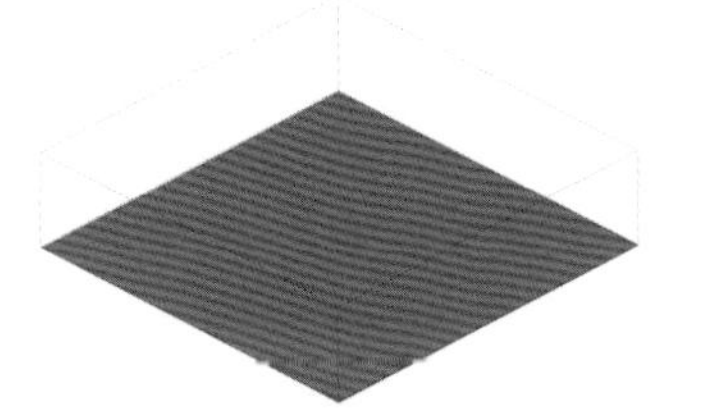

1.
square plan = flexibility

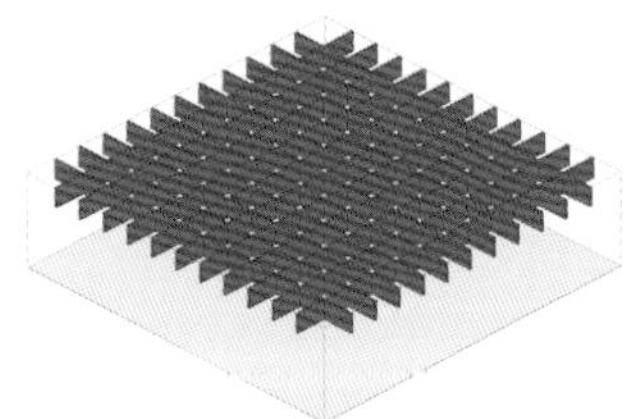

2.
grid structure creating column-free space

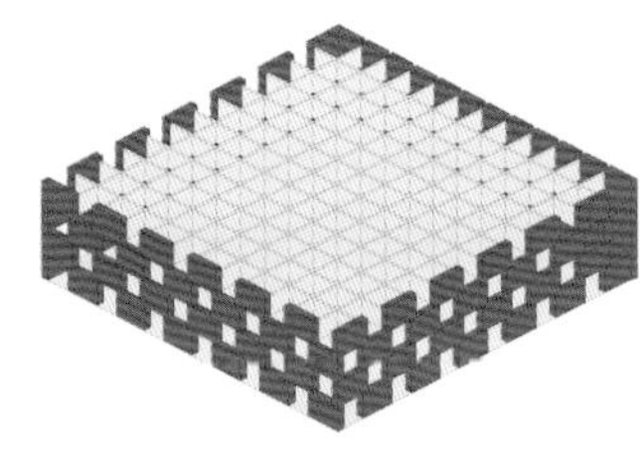

3.
grid transferred to exterior walls

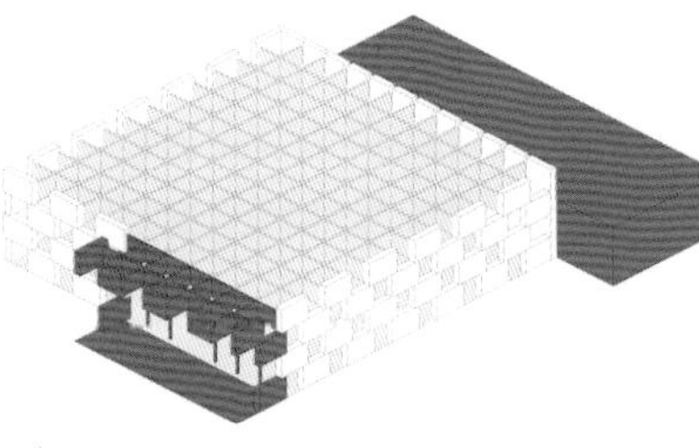

4.
kitchen and entrance added

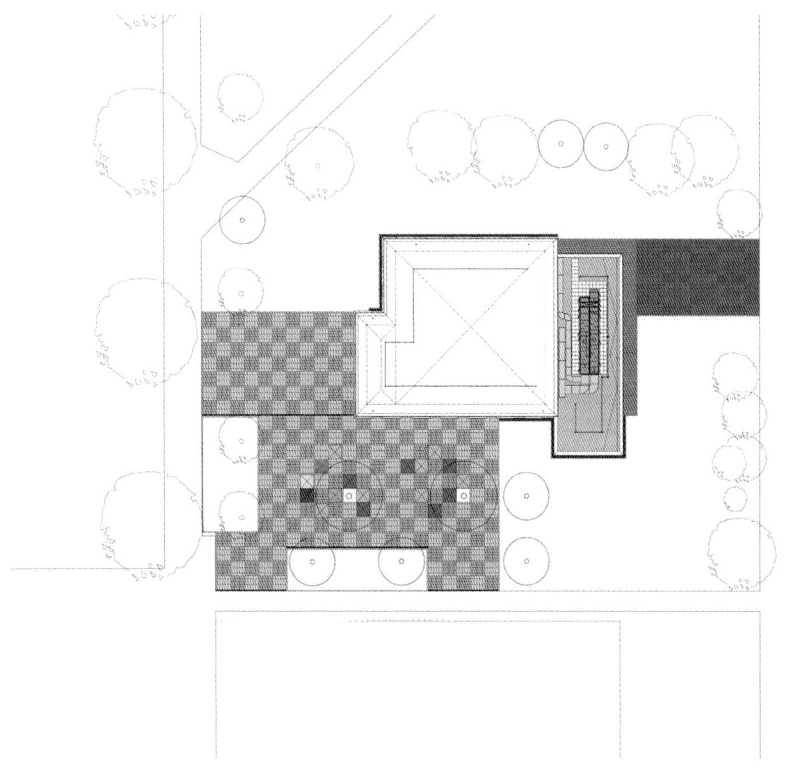

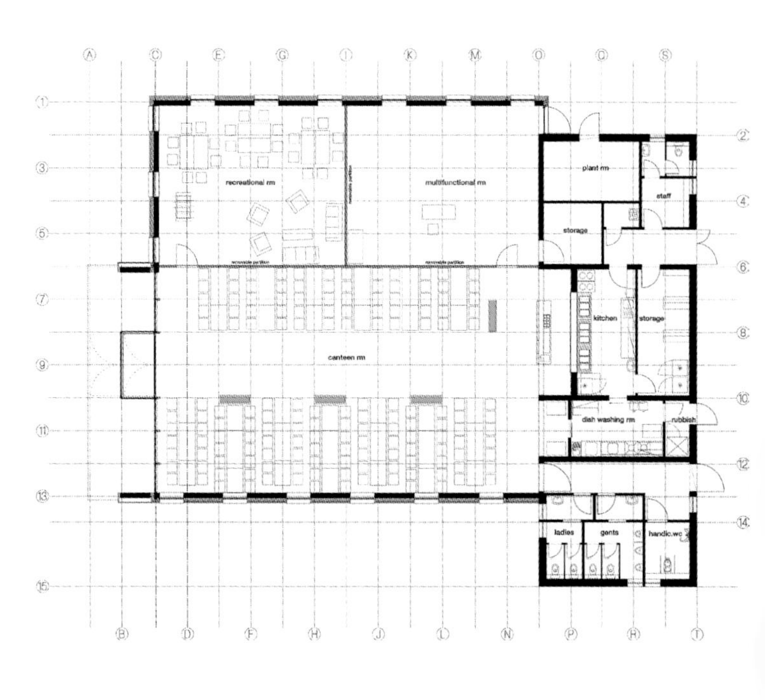
recreational rm
multifunctional rm
plant rm
staff
storage
canteen rm
kitchen
storage
dish washing rm
rubbish
ladies
gents
handic.wc

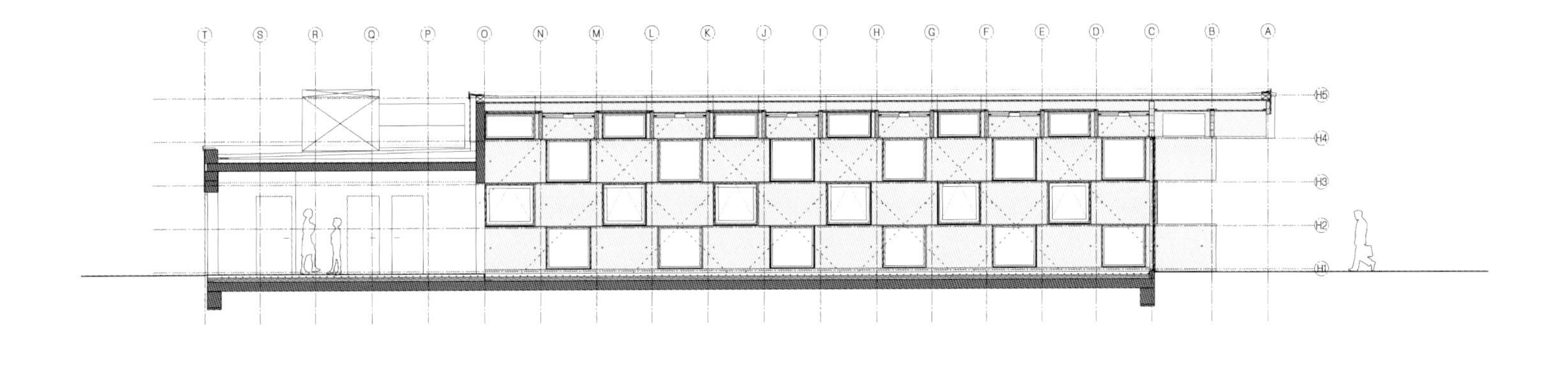

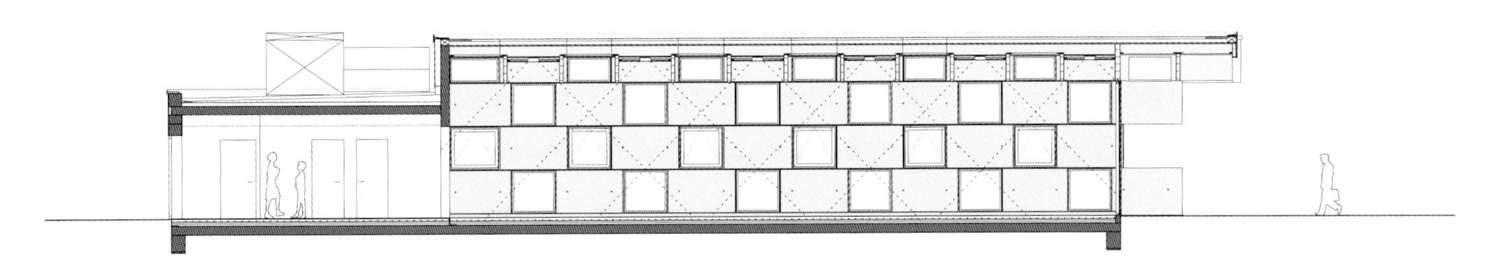

The new school canteen of the Salmtal Secondary School in Germany was designed by SpreierTrenner Architekten as a multifunctional building with the greatest possible flexibility. The checked windows and the bright red facade were inspired by the idea of creating a playful and interactive, yet efficient building. The starting point was the square plan, which allowed for maximum flexibility as it's multidirectional as well as easy to furnish for concerts uses, etc. Then all seats would be relatively close the stage.

To span a square plan most efficiently a two-directional grid was used. The grid got transferred to the facade as well, so the height of the room, the size of the windows and an efficient ratio for the wood trusses determined the grid proportion .

The big glazed entrance opens up the main canteen room to the outside and represents a welcoming gesture. The cantilevering canopy creates a transition zone between the interior and the playground. The roof grid of the main room consists of 10cm thick and 1m high wood trusses.

萨尔姆塔尔中学餐厅是一座多功能的建筑，建筑空间设计较为灵活。格子状的窗户和亮红色的立面设计灵感主要源于这样的想法：打造一处富有趣味性和互动性的高效建筑。项目的出发点是借用方形创造建筑外观，从而使空间更具灵活性。设计师创造了多个空间，其中包括举行音乐会的空间。

为了以最高效的方式实现方形的空间设计，设计师采用了双向式的网格结构。网格结构也体现在立面设计上，因此，立面的规模由以下元素决定：房间高度、窗户的大小及木质桁架的比例。

用玻璃打造的大型入口将餐厅与外部空间联系在一起。悬臂式结构是建筑内部空间与户外操场之间的过渡空间。屋顶的网格结构为 10cm 厚、1m 高的木质桁架。

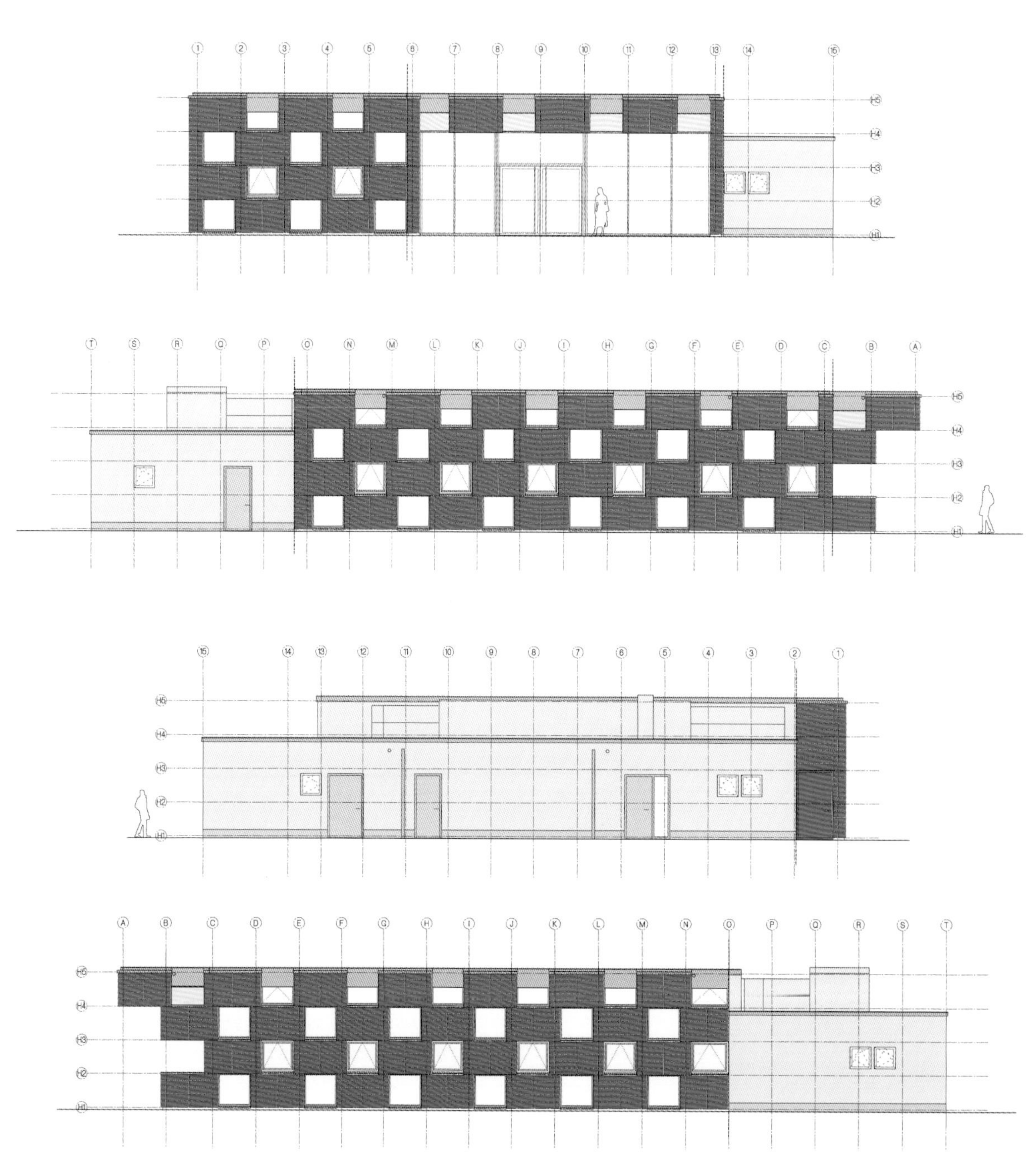

CTA Morgan Station

CTA摩根车站

Design Company / 设计事务所:
Ross Barney Architects
Location / 地点:
USA（美国）
Photography / 摄影:
Kate Joyce Studios

视觉亮点

建筑外立面简洁、醒目，颇具现代时尚感。

Morgan
1000W
200N

Morgan

cta
Morgan
STOP

cta
MORGAN STATION
STOP

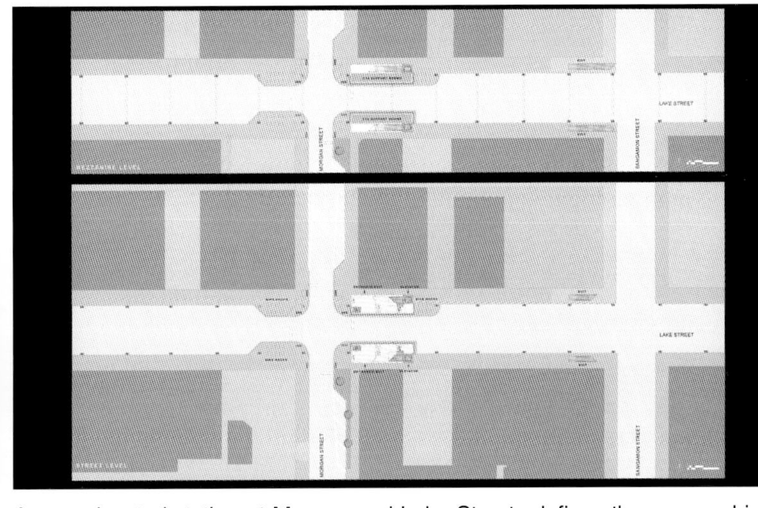

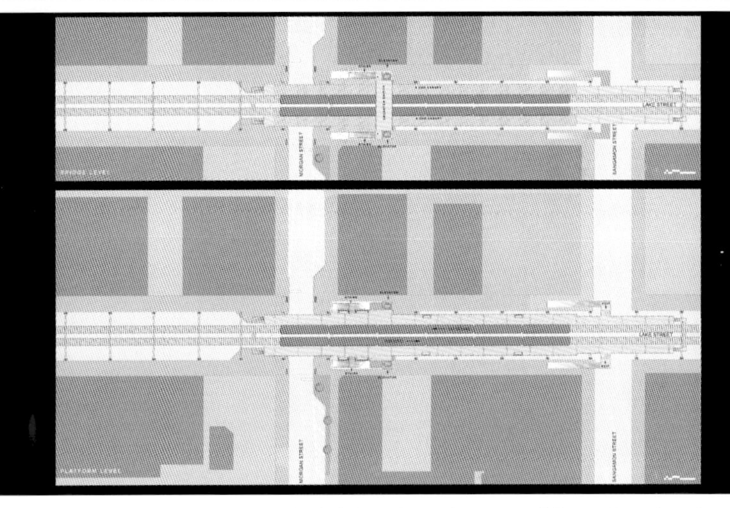

A new elevated station at Morgan and Lake Streets defines the geographic center and character of Chicago's Fulton Market District, an industrial area transformed into a multi-faceted neighborhood with emerging residential and retail uses.

The Market serving the City's wholesale food vendors, is a rich combination of warehouses, off-the-beaten-path restaurants, specialty purveyors, loft conversions and boutique stores. To reinforce this character, material selections take cues from nearby—steel, glass, concrete, polycarbonate, granite and cast iron are all used in adjacent structures.

To maximize station visibility and pedestrian access from the active Randolph Street corridor, stationhouses are located at grade level. New trees, landscaping and artist-commissioned bicycle racks are located along Lake Street to soften the industrial character. Accessibility, durability, and ease of maintenance were prime functional concerns. Each stationhouse has an ADA-compliant elevator that provides disabled passengers access to the platform and to the transfer bridge above between inbound and outbound platforms.

位于芝加哥摩根与雷克大街上的新建车站进一步体现了富尔顿市场区的中心位置和主要特色。随着该地区住宅区和商业零售区的不断涌现，该地区已经从一处工业区转变成一处多姿多彩的住宅和商业区。

该地区拥有各色商业空间，如大商场、特色饭馆、特色食品供应店、精品店，等等。项目设计师从周边建筑设施中汲取了诸多灵感，例如建筑材料选用钢材、玻璃、聚碳酸酯、花岗石、铸铁等。

为了使整座车站更加醒目，并方便人们从繁华的伦道夫大街步行来到这里，车站大厅被设置在整座建筑的一层。新栽的树木、景观绿化带和由艺术家设计的自行车道均沿雷克大街设置，丰富多彩的元素打破了原有地块的工业化特征。设计师的设计重点体现在建筑便于出入、耐久和方便维护等方面。车站大厅均设有自动数据采集式电梯，方便残障人士的通行。

New Talca's University Library

塔尔卡大学图书馆

Design Company / 设计事务所:
Valle Cornejo Architects, José Luis Gajardo Architect
Location / 地点:
Chile（智利）
Area / 面积:
2 951 m^2
Photography / 摄影:
Pablo Riquelme Adams

视觉亮点

建筑一侧立面采用整齐排列的立柱形式，增强了建筑的线性特征，使建筑具有更强的可识别性。

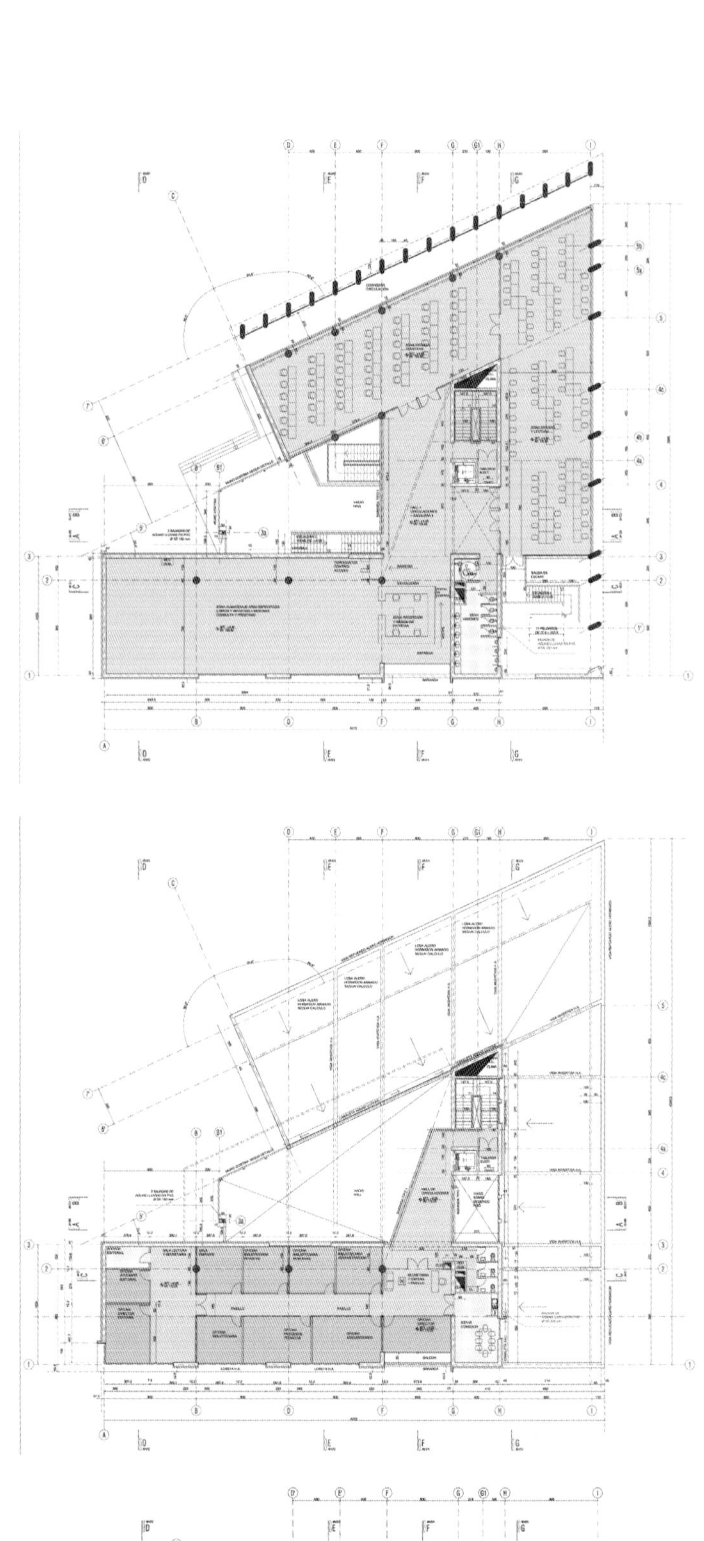

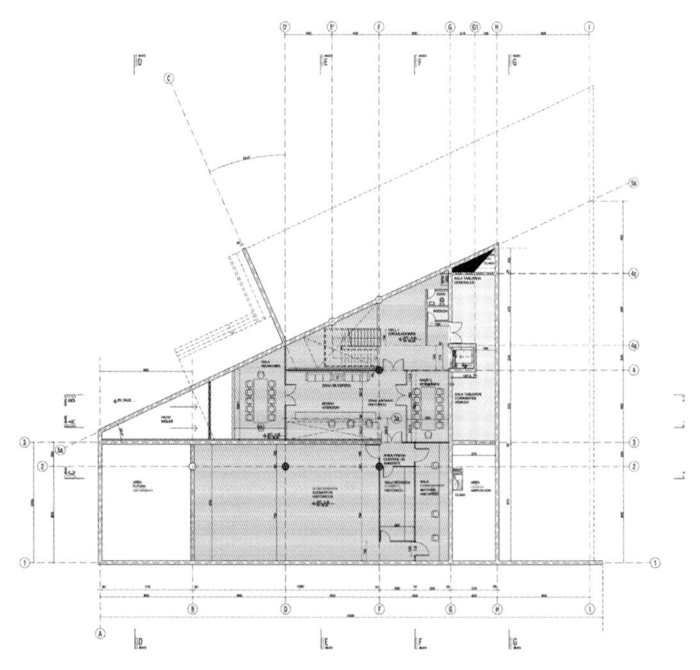

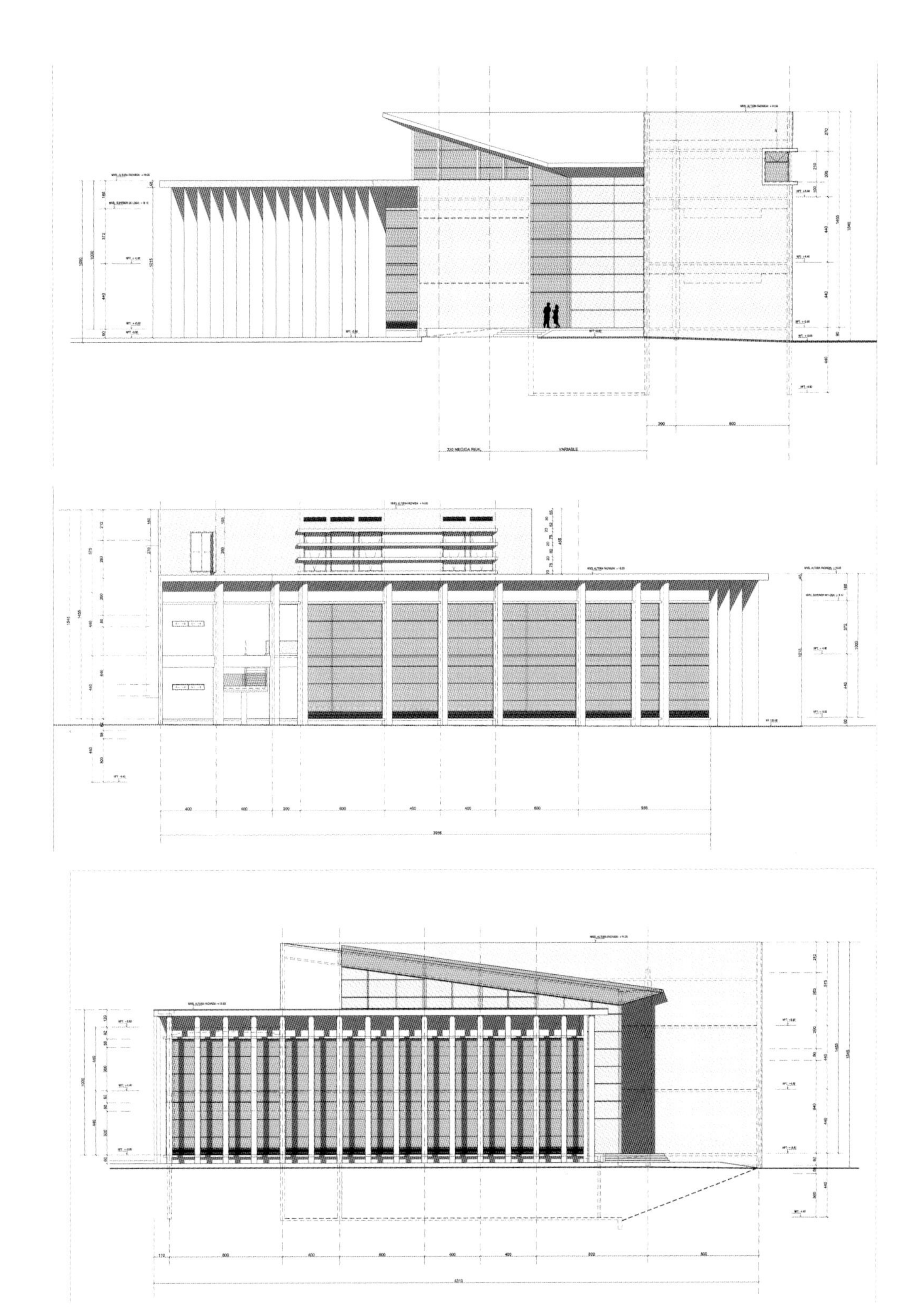

The new library would become a contemporary building, both a manifestation of the University's commitment to energy efficiency and a fundamental symbol of the University's vocation of educating its students.

The new structure serves many functions at the heart of the campus. The Central Library completes the campus system of pedestrian circuits and melds with both the interior and exterior facades of the University buildings, as well as the interior spaces of the library building itself. The building's architecture evokes both the solemnity of its education purpose and the role of a library as a meeting place -- for a library is a place where both ideas and people interact. To achieve this, the design prominently features a grand hall, a refuge where the library's distinct areas come together. The facade design is a series of pillars that open to the exterior north side of the building. The facade both regulates the building's direct sunlight and serves as a recognizable image, a face for the campus and the library. The access point is a large, open space, a lengthening of the main hall that lies parallel to cozier spaces looking west. Inside, the different floors are stacked one on top of the other, in a dynamism that nods at modernity with its privileging of angles and geometric forms.

塔尔卡大学新建图书馆将是一处富有现代特色的建筑，既能展现大学节约能源的决心，又能体现大学教书育人的特色。

位于校园中心的这处新建图书馆具有多重功能。图书馆与校园中人行通道相连接。建筑的内部、外部空间完美相融，图书馆建筑的内部空间设计与外立面风格保持一致。该建筑在设计上不仅展现了教育场所的神圣性，也展现了图书馆作为会面场所的空间角色——既可供人们交流思想，又方便人们会面交谈。为了满足建筑的使用功能，设计师在图书馆内部设计了开阔的大厅。图书馆建筑立面采用一系列的立柱设计，而建筑北立面则采用开放式的设计形式。这样的立面不仅使整座建筑拥有良好的光照条件，能够接受直接的自然光照，更突出图书馆的外观特点，使其更容易被识别。建筑的入口处为一处开阔的开放式空间。该入口处是内部大厅的延长部分。

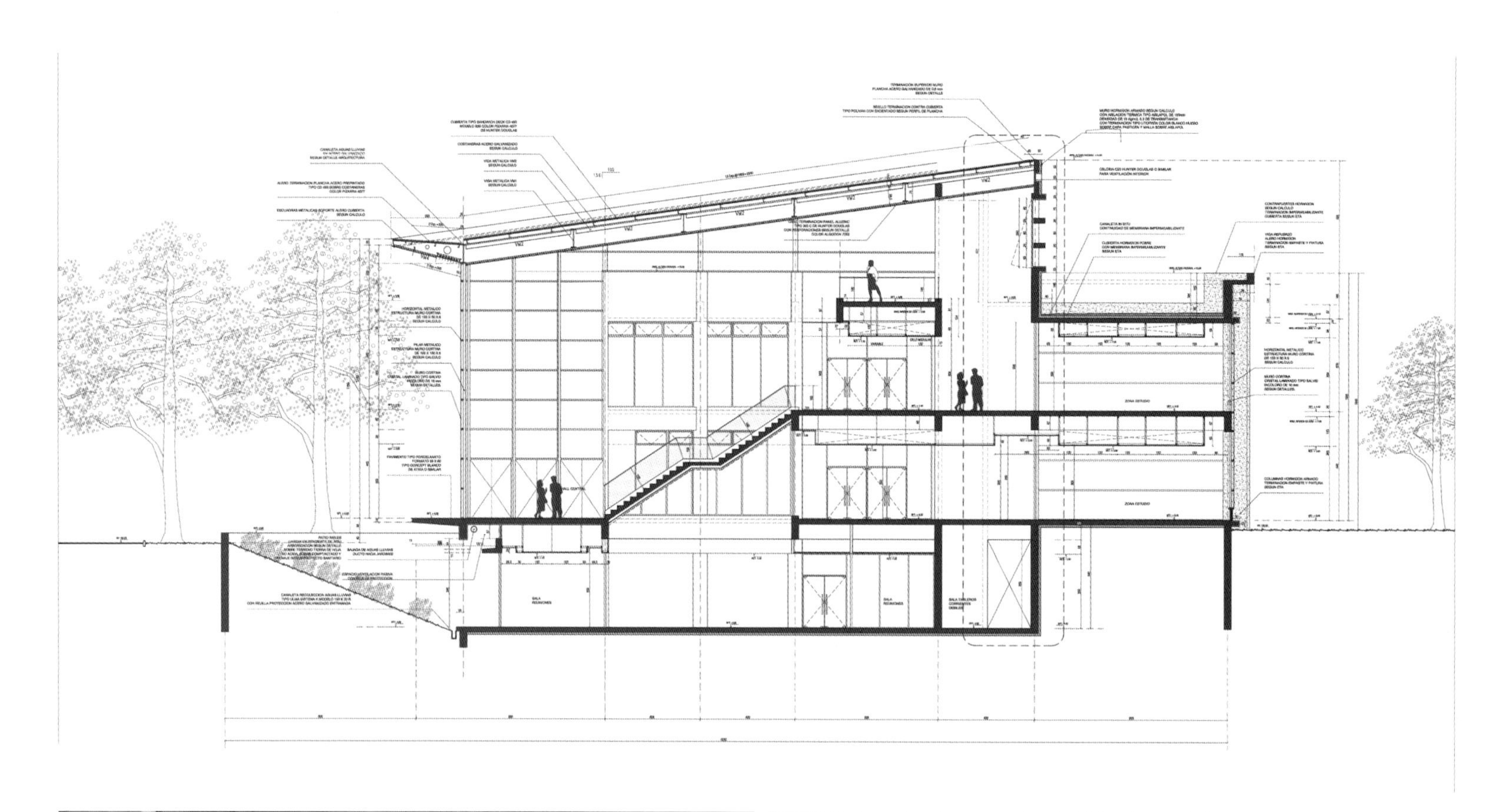

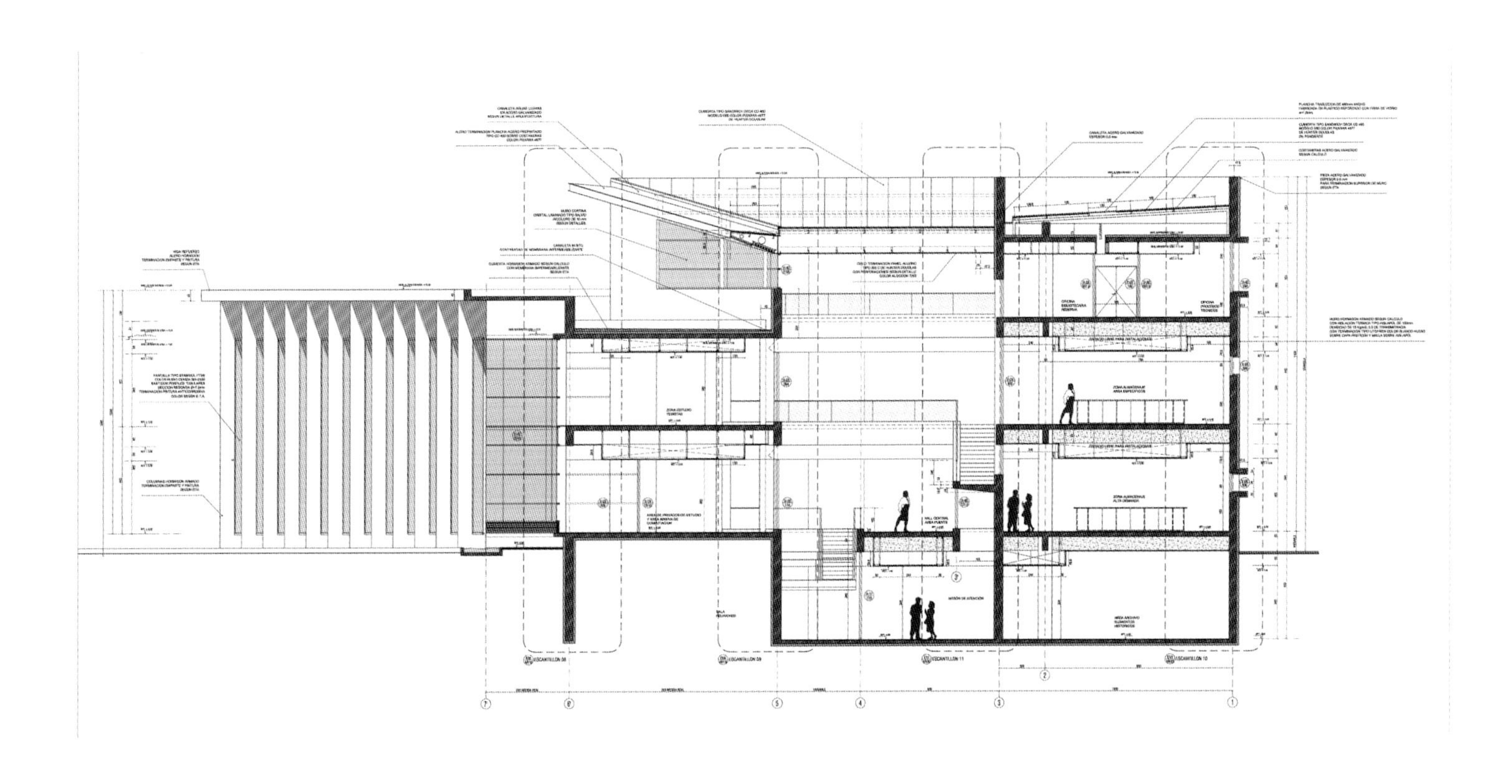

Red Apple Apartment Building

红苹果公寓

Design Company / 设计事务所:
Aedes Studio
Location / 地点:
Bulgaria（保加利亚）

视觉亮点

长方形元素或凸出或凹陷，创造了一个整齐又富有动感的建筑立面。

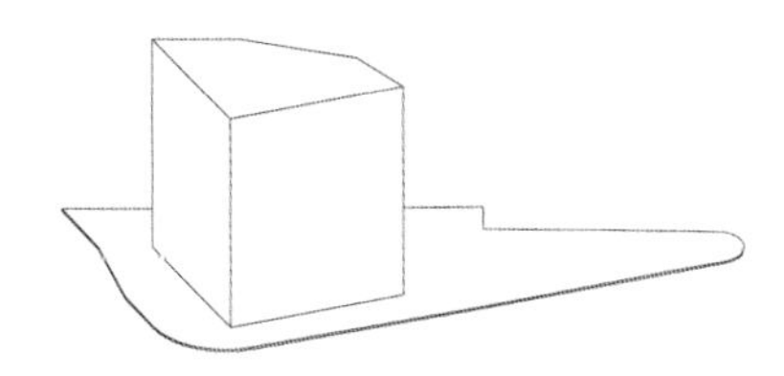
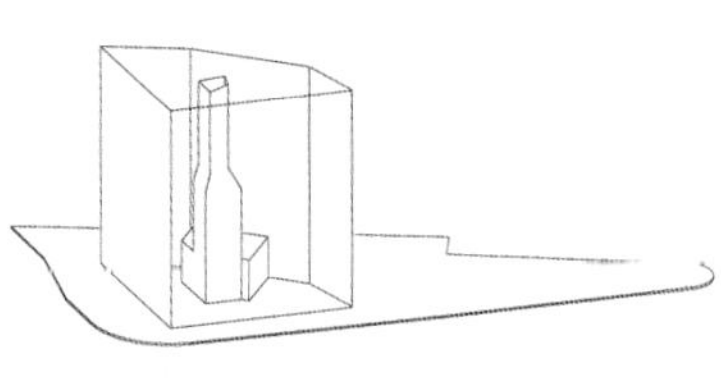
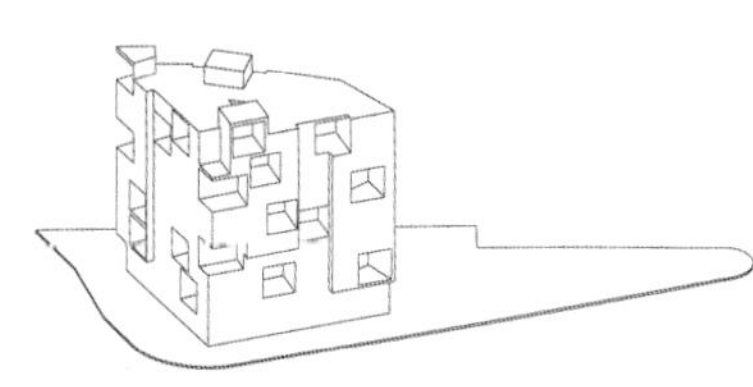
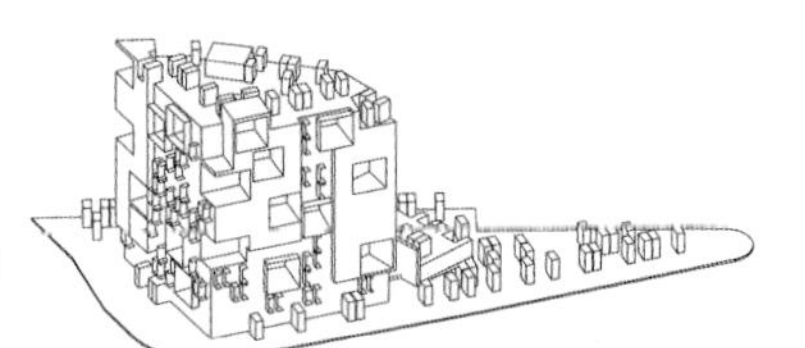

The surrounding neighborhood consists mostly of apartment blocks that date back from the 70's. The buildings are large with enough space in-between and plenty of greenery. Because the whole area is built in relatively short period of time and not very long ago, it lacks the typical historic layers of the city center. Here the connection to nature is direct enough, the access to all city-conveniences – fast enough and easy, what makes the area nice to dwell.

The code of the building consists of a perforated brick shell. Its outline deliberately follows the irregularities of the site, creating acute angles which enhance the perspective and establish ▯ significant character. The openings are completely similar, in strict order that originates from the brick's grid. In random places they are missing or are replaced by large break-throughs (two storey windows).

The apartments contain island-like situated volumes (rooms) away from the façade. This makes it possible for the inhabitants to notice the rhythm of the façade-openings from the inside.

The "old", "abandoned" and once again discovered Living factory is an environment rich in different aspects. It unifies the advantages of the contemporaneity as well as the past. This way the building doesn't just fit it's surroundings – it delivers what the neighborhood lacks – history, past and memories.

该项目的周边街区建筑主要是建于 20 世纪 70 年代的公寓楼。各个建筑之间的空间很大，拥有大片的绿地。整个区域建成时间距离现在时间不长，故而缺少市中心的历史色彩，而这也是该项目要应对的设计重点所在。建筑与自然之间拥有直接的联系，通往所有城市便利设施也很便捷。

整座建筑拥有穿孔式砖墙结构。其轮廓与不规则的地块外观相呼应，赋予整座建筑与众不同的外观。建筑的窗口设置在砖墙上，以统一的外观遵循严格的设计秩序呈现出来。在几处地方，该有的窗口消失了，代之以凸出的两个楼层高的大型窗口。

公寓楼内拥有岛屿式的房间设置，距离立面有一段距离。这就使得生活其中的人能够从室内观赏到富有节奏感的立面窗口设置。

公寓周边那座“古老的”“废弃的”又被重现发现的厂房富有丰富多彩的侧面。其将现代特色与历史感充分衔接起来。公寓建筑不仅与周边环境相适应，还展示了周边街区所缺失的一些东西——历史感和人们对过去的回忆。

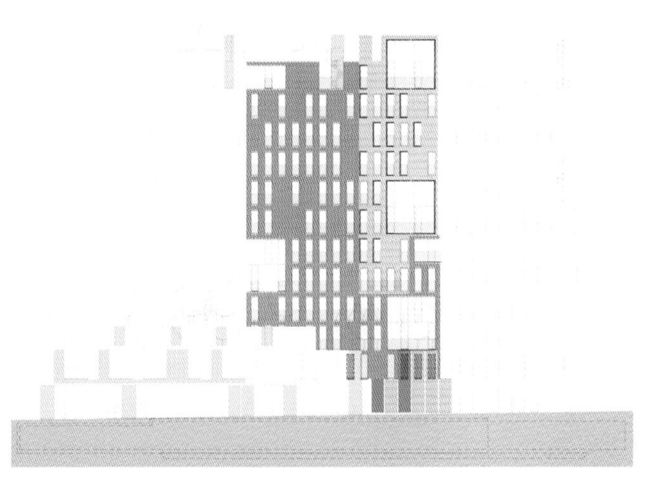

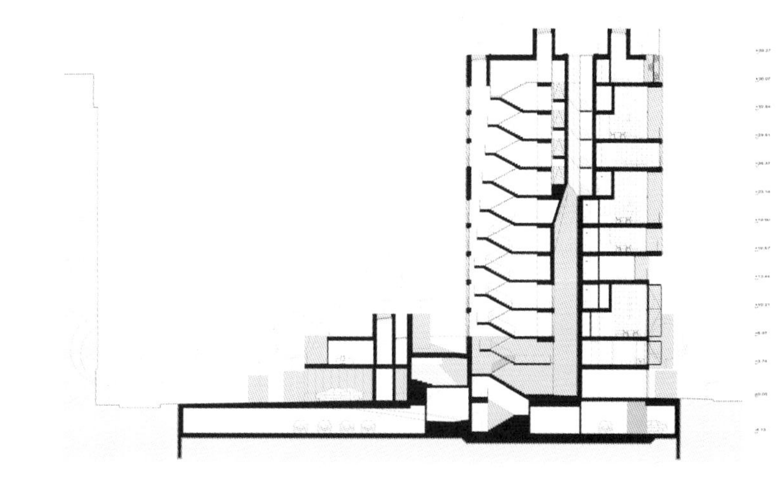

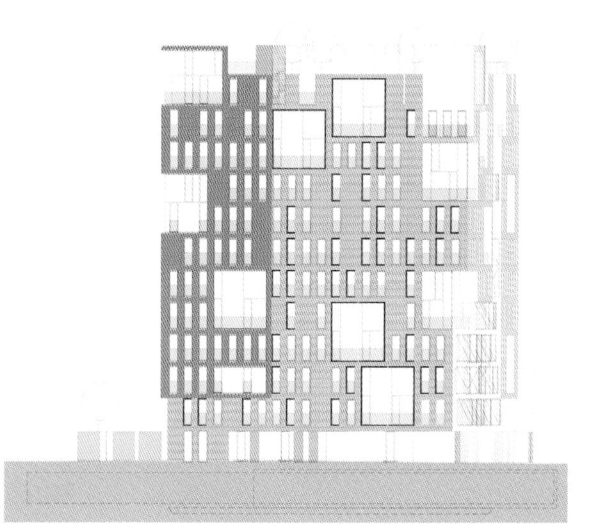

Bayuquan Library

鲅鱼圈图书馆

Design Company / 设计事务所:
上海都设建筑设计有限公司
Project Architect / 设计师:
凌克戈
Location / 地点:
China（中国）
Area / 面积:
site 700000 m^2, buliding 10000 m^2

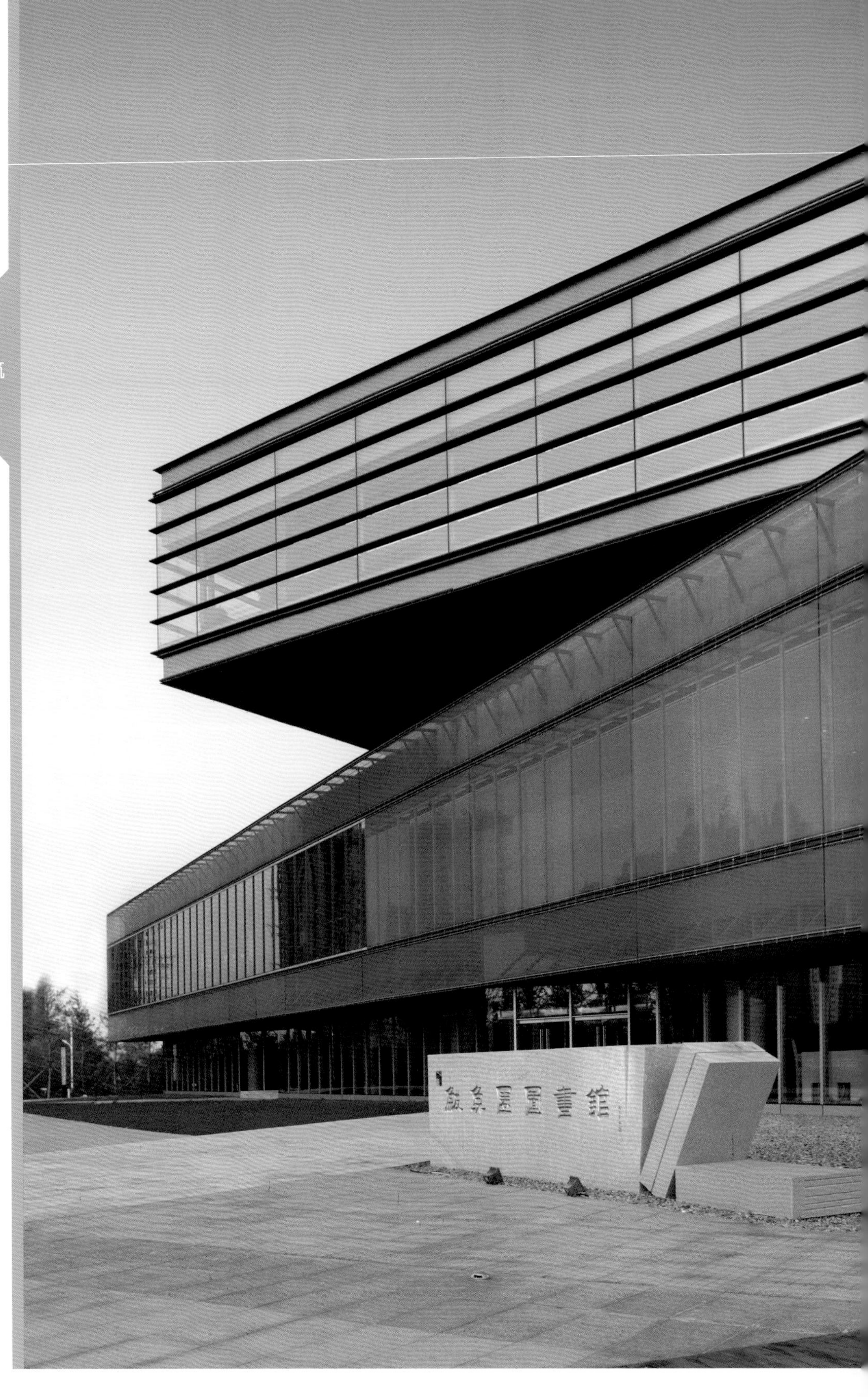

视觉亮点

两个扭转相叠的矩形体块，就像两本交错叠放的书，恰如其分地表现出建筑的使用功能。

ZOOMLION

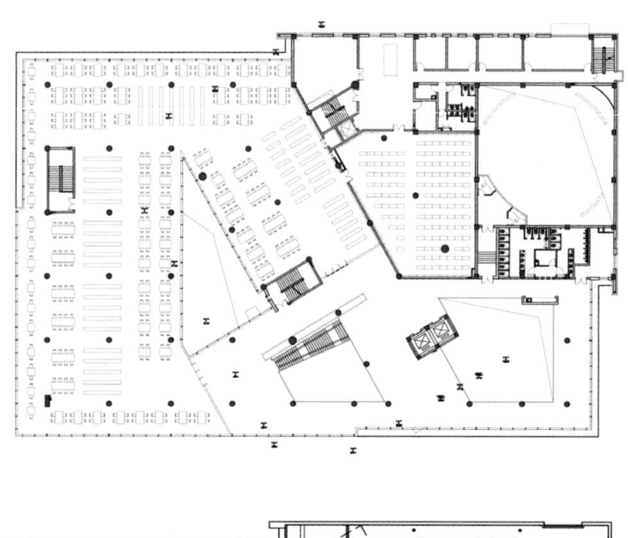

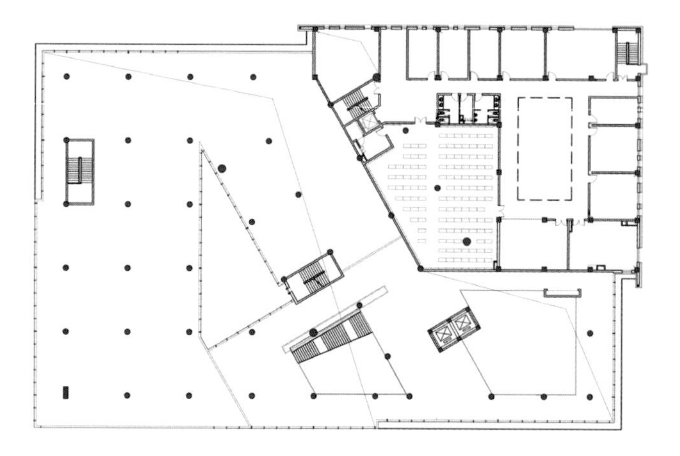

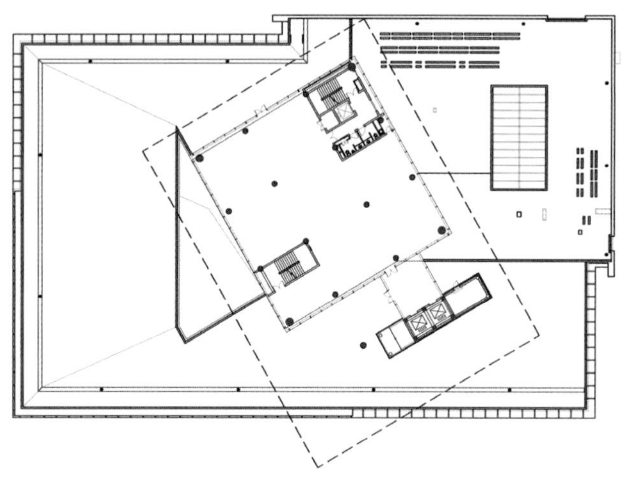

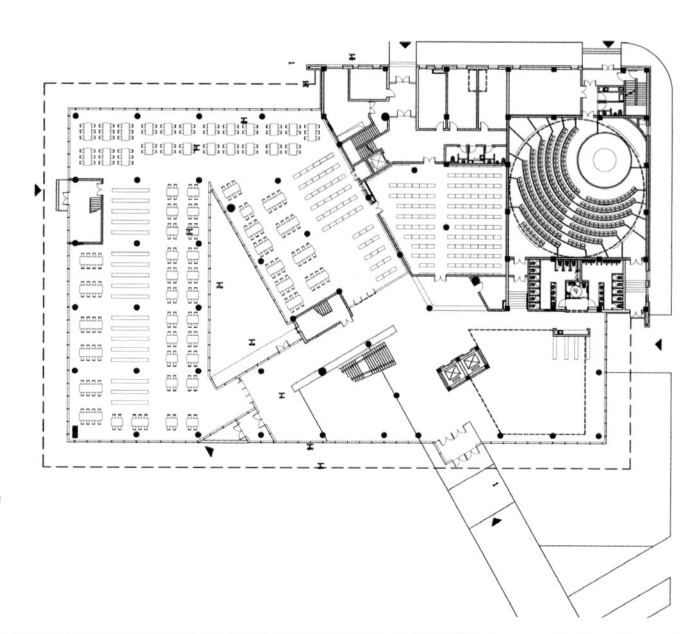

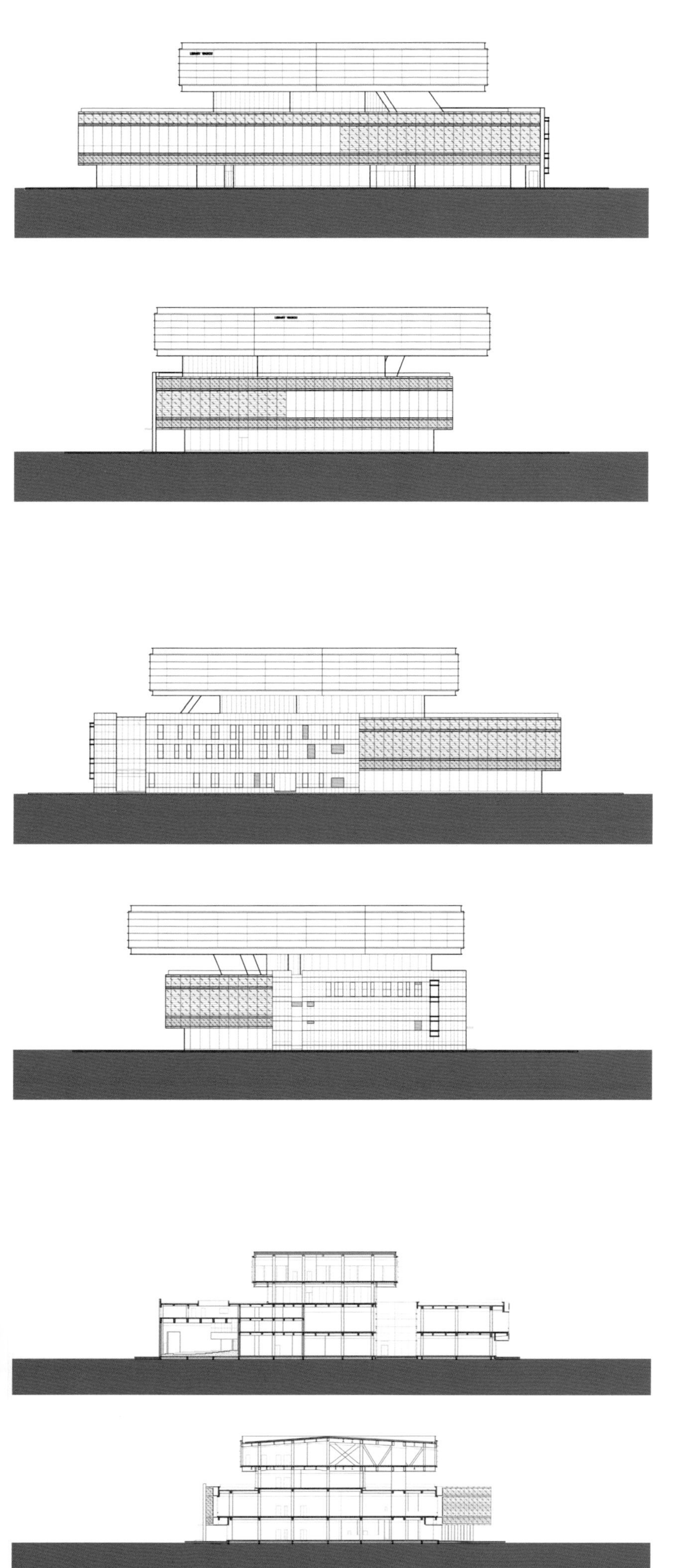

Bayuquan library is another project designed by DSD in Bayuquan near Bayuquan Theatre. The idea of this design was from two rotated books. And the rotated blocks provide a best view. The library is mainly made up of 3 parts including the basement block containing 2 levels, the cantilever (3rd floor) and the connection. The roof of the lower block provide exterior space for outdoor activities. In order to have a better view inside, the cantilever employs a full curtain wall. Additionally, the triangular atrium ensures enough daylight of the reading room.
The library use the similar material and construction strategy as the theatre.

图书馆采用两个扭转相叠的矩形体块，寓意两本交错叠放的书。扭转的体块具有最佳的观景视野，巨大的出挑结构塑造出极富结构美感的现代建筑形象。图书馆分为基座（一、二层）、悬挑结构（四层）以及结合部分（三层）。基座屋顶可为三楼提供室外活动场地，图书馆所有的对外部分包括阅览室都可以看到二道河公园。图书馆基座部分采用与大剧院相呼应的材料及构造方式，悬挑的部分采用玻璃幕墙结构，使其具有270° 的极佳观景视野，并成为图书馆的标志。
图书馆阅览中设置了一个三角形的庭院，为阅览室提供阳光和绿化。图书馆拥有一个报告厅，供内部学术研讨及新书发布等使用。

地方文献借阅区

叠式建筑

叠式建筑指的是结构相同或相近的几何体块在水平方向或垂直方向上的层层叠加。叠式的结构更容易营造视觉上的连续感，使建筑外观更富节奏感和韵律感。

Library of Birmingham

伯明翰图书馆

Design Company / 设计事务所:
Mecanoo
Location / 地点:
UK（英国）
Area / 面积:
35 000 m^2
Photography / 摄影:
Christian Richters

视觉亮点

建筑外墙采用金银细丝式的设计形式， 不仅体现了建筑所在城市的工业文化，更建造了一个新颖别致的具有质感的层叠式建筑。

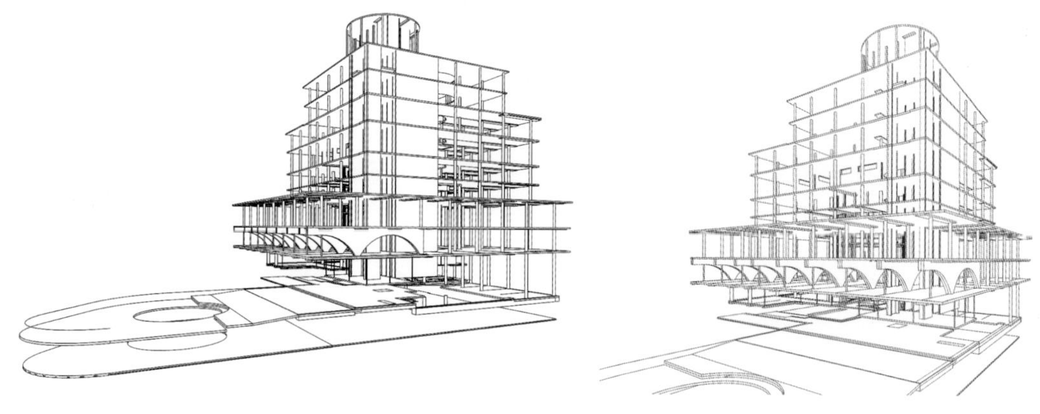

THE REP

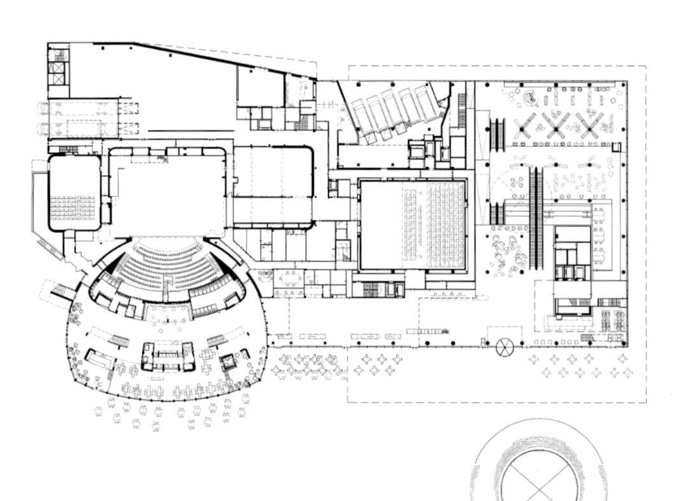

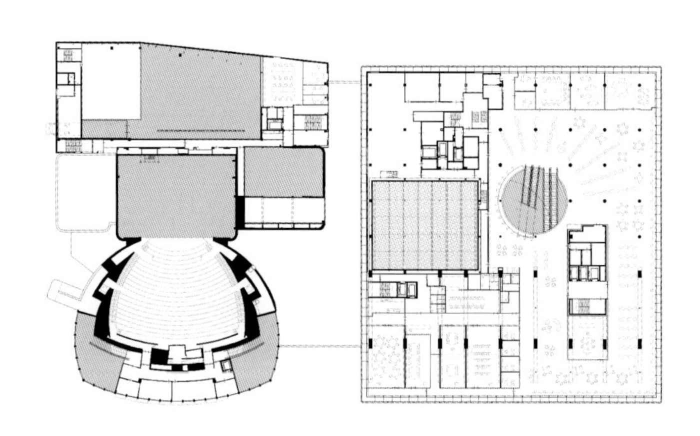

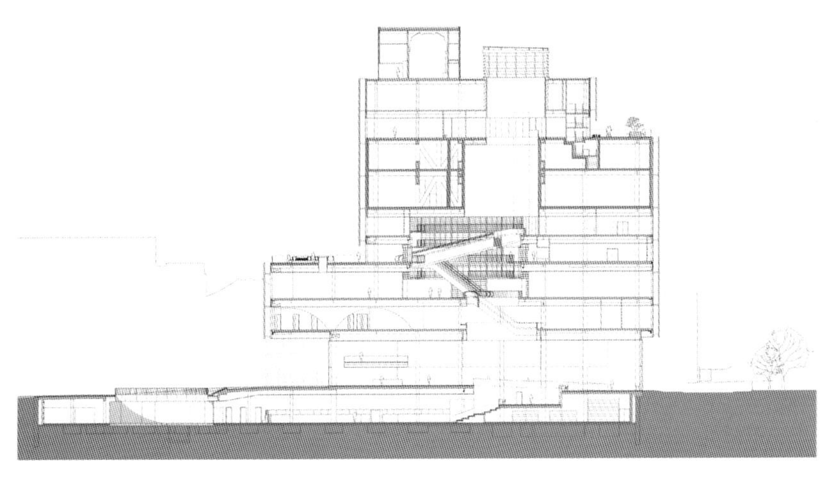

Centenary Square, the largest public square in the heart of Birmingham, currently lacks cohesion or a clear identity or atmosphere. Mecanoo's design transforms the square into one with three distinct realms: monumental, cultural and entertainment. These palazzos form an urban narrative of important periods in the history of the city; The Repertory Theatre (REP), a 1960s concrete building, the Library of Birmingham, designed in 2009 and Baskerville House, a listed sandstone building designed in 1936. The Library of Birmingham is a transparent glass building. Its delicate filigree skin is inspired by the artisan tradition of this once industrial city. Travelators and escalators dynamically placed in the heart of the library forms connections between the eight circular spaces within the building.

百年广场是伯明翰市中心最大的公共广场，然而，这处广场缺乏凝聚力和独特的个性，或者说缺少某种氛围。该项目设计要突显这座广场三个主要特色：纪念性、文化性、娱乐性。该项目将这座城市的三个重要历史时期串联到了一起：剧院（即 REP）为建于 20 世纪 60 年代的混凝土建筑，伯明翰图书馆于 2009 年设计完成，巴斯克维尔为设计于 1936 年的砂岩建筑。图书馆的悬臂部分不仅是一处大型的华盖，为伯明翰图书馆和剧院的共用入口提供遮蔽，还是一处大型的城市观景台，使人们可以一览广场上的风景。伯明翰图书馆为一处通透式的玻璃建筑。其精致的金银细丝外表面设计灵感源于这座工业城市传统的工匠形式。位于图书馆中心的移动步道和自动扶梯将建筑内的八层空间联系在一起。

REP

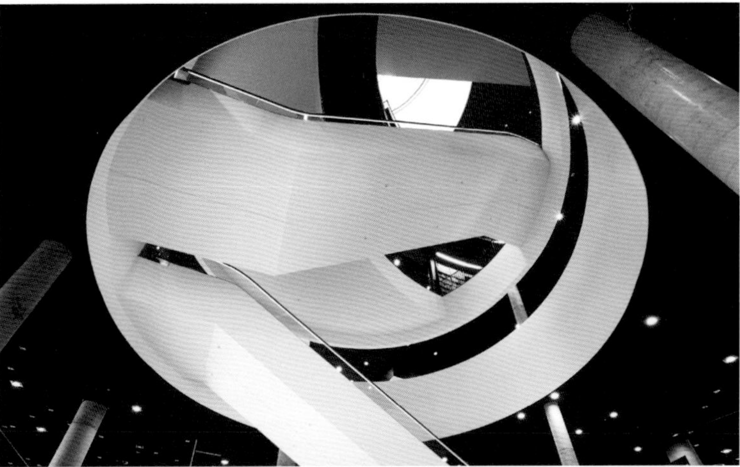

叠式建筑

文化建筑

Kunstcluster, Nieuwegein

新维根艺术中心

Design Company / 设计事务所:
van Dongen-Koschuch
(former employees of the Architeken Cie)
Project Architect / 设计师:
Frits van Dongen & Patrick Koschuch
Location / 地点:
The Netherlands（荷兰）
Area / 面积:
27 550 m^2
Photography / 摄影:
Allard van der Hoek

视觉亮点

建筑一层层“晶莹剔透”的结构和充满绿色生机的停车场，定会给人们留下深刻的印象。

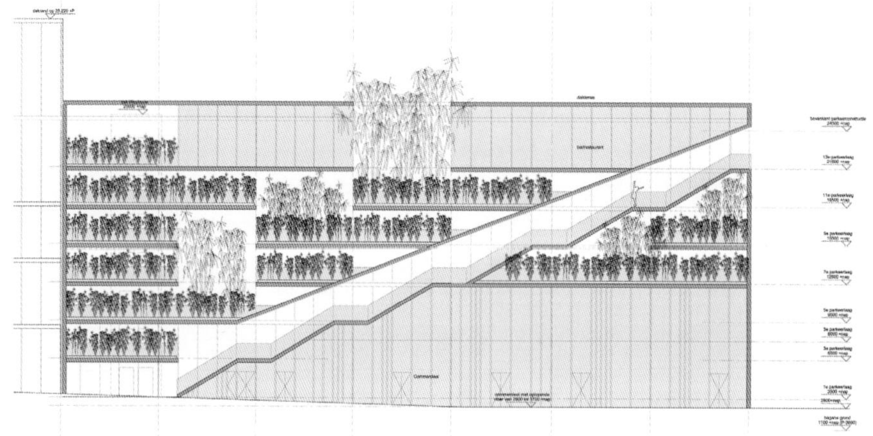

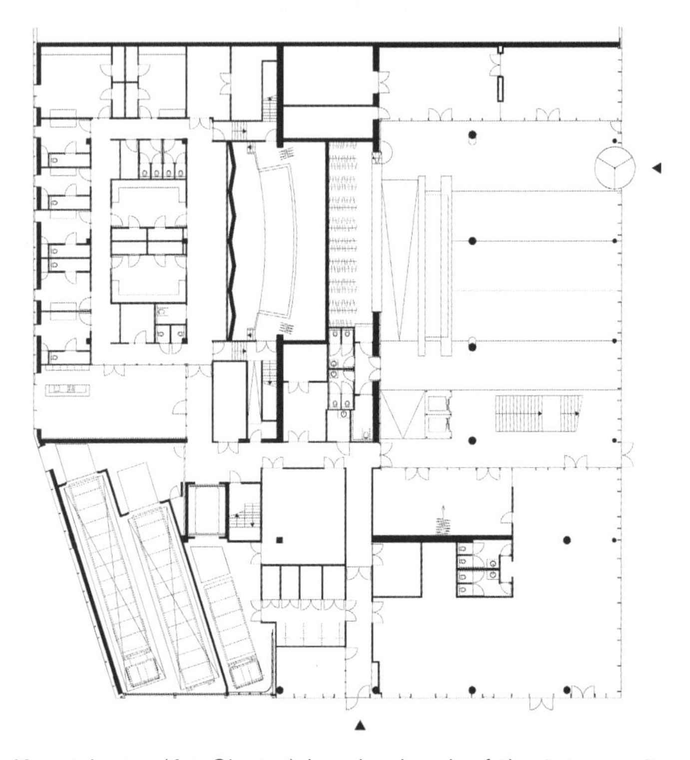

The Kunstcluster (Art Cluster) is a landmark of the town centre of Nieuwegein, which is to be redeveloped over the next few years. A new town hall is to be built, with shops, homes and offices above ground, and under-ground parking. The site of the Kunstcluster consists of two conjoined blocks: the theatre with arts centre as well as the multi-storey car park combined with retail space.

The glass façde is printed, to give the illusion of (stage) curtains. This applied print also considerably reduces the percentage of light penetration via the south and east façde, preventing overheating on sunny days. The foyers, and therefore most people, are located on the Stadsplein façde to the south. Here, the pigment of the print is sufficiently transparent and translucent to make the image visible from inside to outside, and from outside looking in.

In the redeveloped town centre nearly all cars will go underground. The multi-storey car park next to the Kunstcluster is an exception. This block is literally a green lung in the stone-built urban environment. The garage frontage on the Stadsplein side is four metres thick. Here a cascade stairway wends its way up/down through a bamboo plantation six metres high. The ground floor accommodates retail, resulting in a living streetscape.

该艺术中心是新维根市中心的地标性建筑，在接下来几年将会有新的扩建内容。这里会建新的市政大厅、商店、住房、办公区，以及地下停车场。该艺术中心包含两处连在一起的街区：设有艺术中心的剧院及多层停车场（含零售空间）。

建筑的玻璃立面被打造出特别的花样，看上去就像是剧院舞台的屏幕。这样的立面设计大大减少了通过东、南立面照射进室内空间的阳光，从而可以避免在阳光充足的日子里室内积聚过多的热量。南部立面使用了通透式的材料，当人们站在南部立面内的室内空间中时，就能一览无余地欣赏到室外的景色。

在这处改造的市中心地区，几乎所有的停车场都位于地下。而靠近新维根艺术中心的多层停车场却是个例外。这处停车场实际上就是钢筋水泥打造的城市环境中的一个“绿肺”。停车场靠近街道的正面墙体厚度为4m，墙体以绿色植物作为装扮。停车场一层还设置了零售区，打造出富有活力的街景。

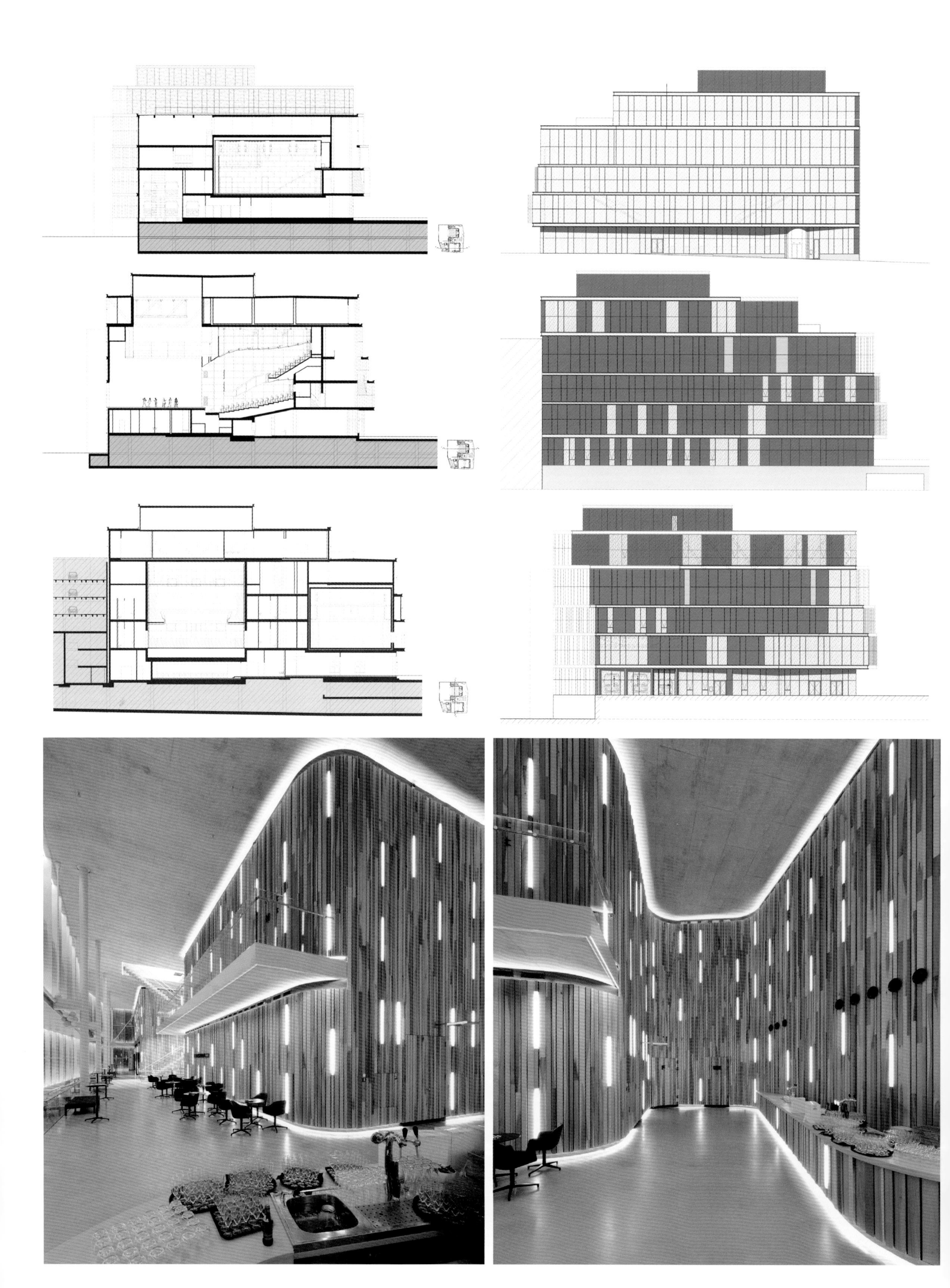

KLEINE ZAAL

叠式建筑

商业建筑

Hotel Lone
龙尔酒店

Design Company / 设计事务所:
3LHD
Project Architect / 设计师:
Silvije Novak, Tatjana Grozdanić Begović, Marko Dabrović, Saša Begović, Ljiljana Đorđević, Ines Vlahović, Željko Mohorović, Krunoslav Szorsen, Nives Krsnik Rister, Dijana Vandekar, Tomislav Soldo, Ana Deg
Location / 地点:
Croatia（克罗地亚）
Area / 面积:
22 157 m²
Photography / 摄影:
3LHD, Cat Vinton, Damir Fabijanic

视觉亮点

流畅的曲线围合出一层层的建筑结构。建筑造型飘逸、潇洒。

URBAN SCHEME
MAIN ENTRANCE
SERVICE
HOTEL LONE
HOTEL EDEN
SEA VIEW

VERTICAL LOBBY SCHEME
FOREST VIEW
POOL VIEW
MAIN ENTRANCE

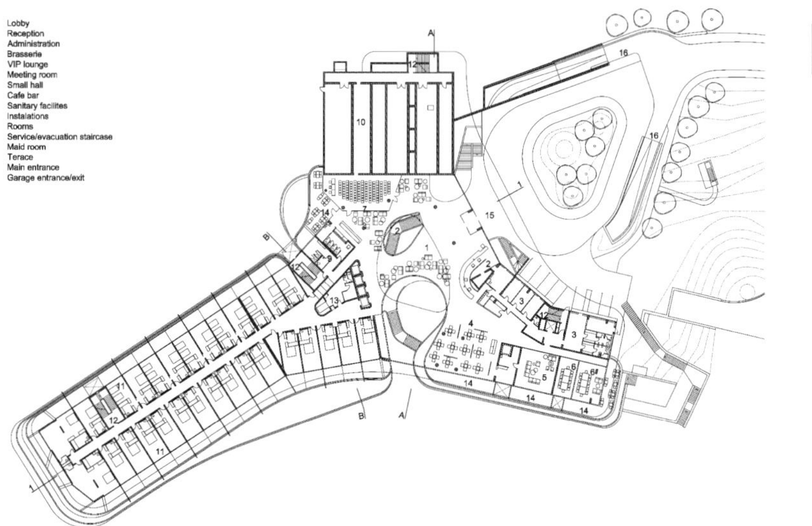

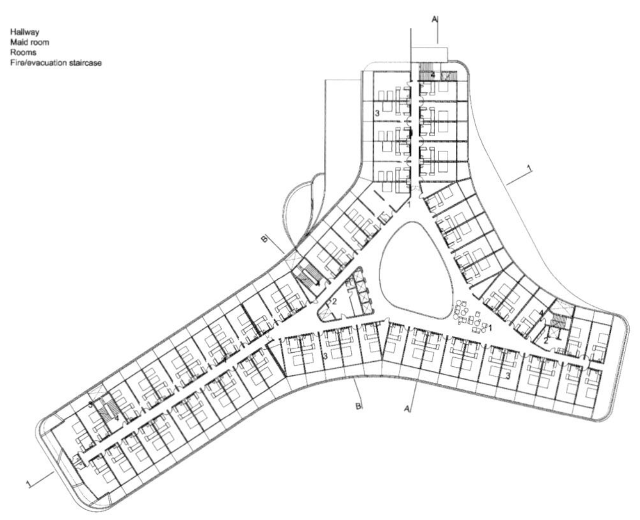

Hotel Lone, the first design hotel in Croatia, is situated in the Monte Mulini forest park, Rovinj's most attractive tourist zone, located in the immediate vicinity of the legendary Eden Hotel and the new Monte Mulini hotel. The surrounding grounds and parkland is a unique and protected region of the Monte Mulini forest on the Lone Bay. The term design hotel is meant to illustrate this as a space that nurtures the concept of an interesting and functional design. Created by a team of renowned Croatian creatives comprised of a new generation of architects, conceptual artists, product, fashion and graphic designers. In addition to the overall architecture, the interiors and the furniture were designed and chosen especially for the hotel in order to achieve a distinct and recognizable identity.

The hotel's identity is recognized through the external design of the building, with a facade that is defined by dominant horizontal lines – terrace guards designed to evoke the image of slanted boat decks. The building's floor plates contract from level to level going up, creating an elevation that is tapered at all angles. The site's complex terrain with dramatic altitude changes determined the locations of internal facility spaces through a dynamic interweaving of public areas and guest suites at all levels. The specific Y shaped ground plan enabled a: rational and functional organizational scheme; quality views from all rooms; and the grouping of public facilities around a central vertical lobby. The main lobby connects common spaces on all levels, creating a central volume of impressive height and scale with interesting views in and around where all vital functions of the hotel take place.

The conceptual assumptions used in the design of the hotel and its interior show evidence of a deep respect towards the achievements of hotel architecture on the Adriatic Coast from the previous century, combining it with a strong modernity expressed primarily in materials, functions and typologies and consequently in architectural forms.

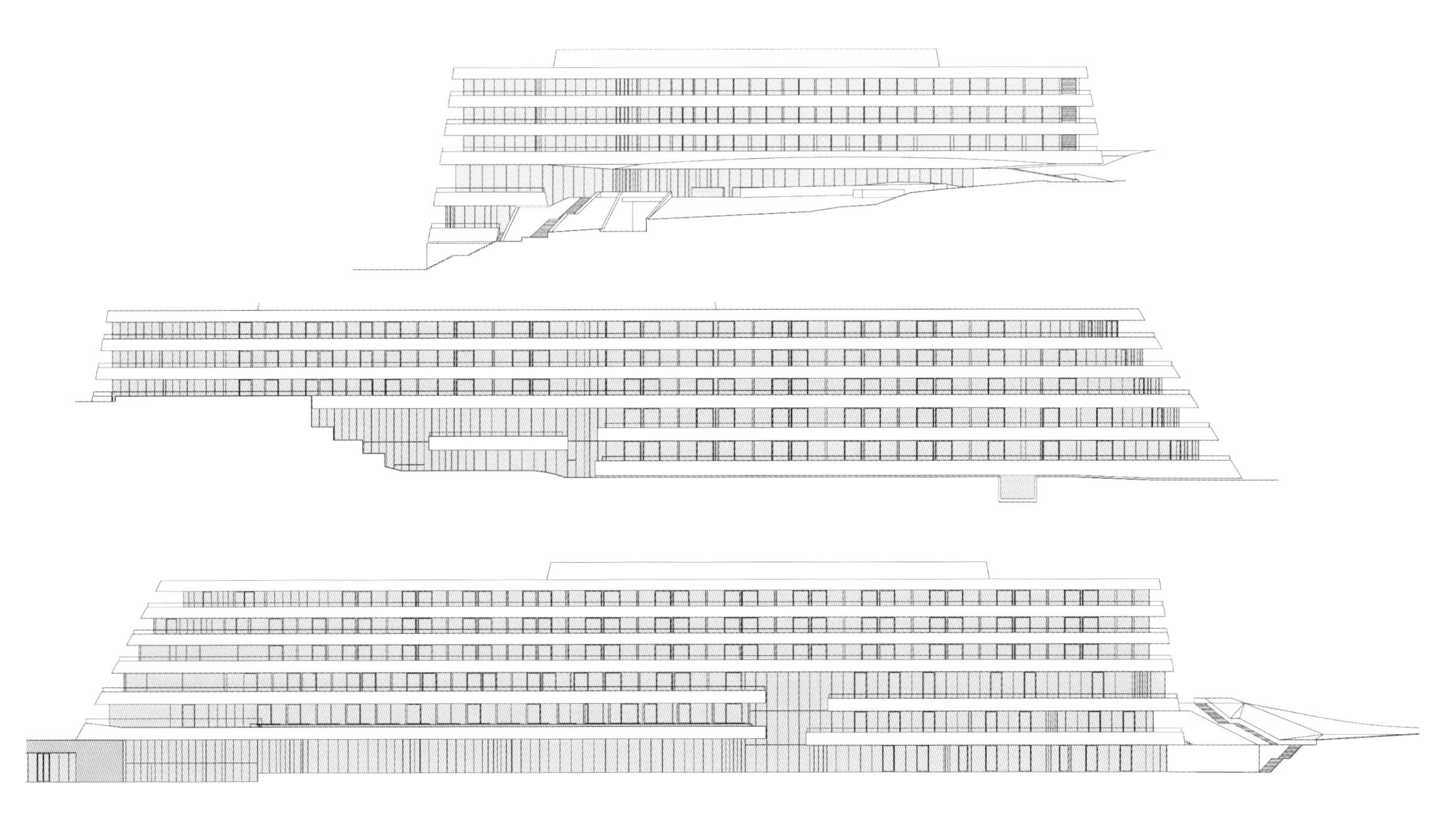

龙尔酒店是克罗地亚首座被设计的酒店，坐落于罗维尼最美丽的旅游景区穆里尼山森林公园内。其周边的环境独特，是龙尔湾穆里尼山保护区。该项目设计师致力于将该酒店打造成兼备趣味性和功能性的概念空间。设计团队由克罗地亚著名的艺术创作人员组成，包括新生代建筑设计师、概念派艺术家，以及产品、时尚和视觉设计师。除了建筑的主体结构外，建筑的室内与家具也是特别为酒店而设计的，从而让该酒店拥有独特的风格。

该建筑外观的设计是酒店最主要的特色，外立面呈现出强烈的线条感，阳台的设计让人联想到倾斜的救生艇的甲板。随着楼层的上升，楼板逐渐往后退让，使得从任何角度看，该建筑都是一个锥形体。由于建筑的地形复杂，竖向变化丰富，每一层的室内公共空间和客房必须呈动态交叉式布局。Y 形的平面布局实现了合理的、能够满足功能需求的组织体系，保证了每个房间的景观视野，并将公共设施围绕着中央大厅布置。中央大厅联系着每一层的公共空间，其极高的高度和空间规模使得人们能够观看中央大厅中及周边的功能区的活动，带给人们无限的乐趣。

该酒店建筑和室内的设计概念充分表现了设计师对 20 世纪亚得里亚海岸建筑卓越成就的尊敬。设计中融入了极强的现代感，这主要体现在建筑的材料、功能和形式设计方面。

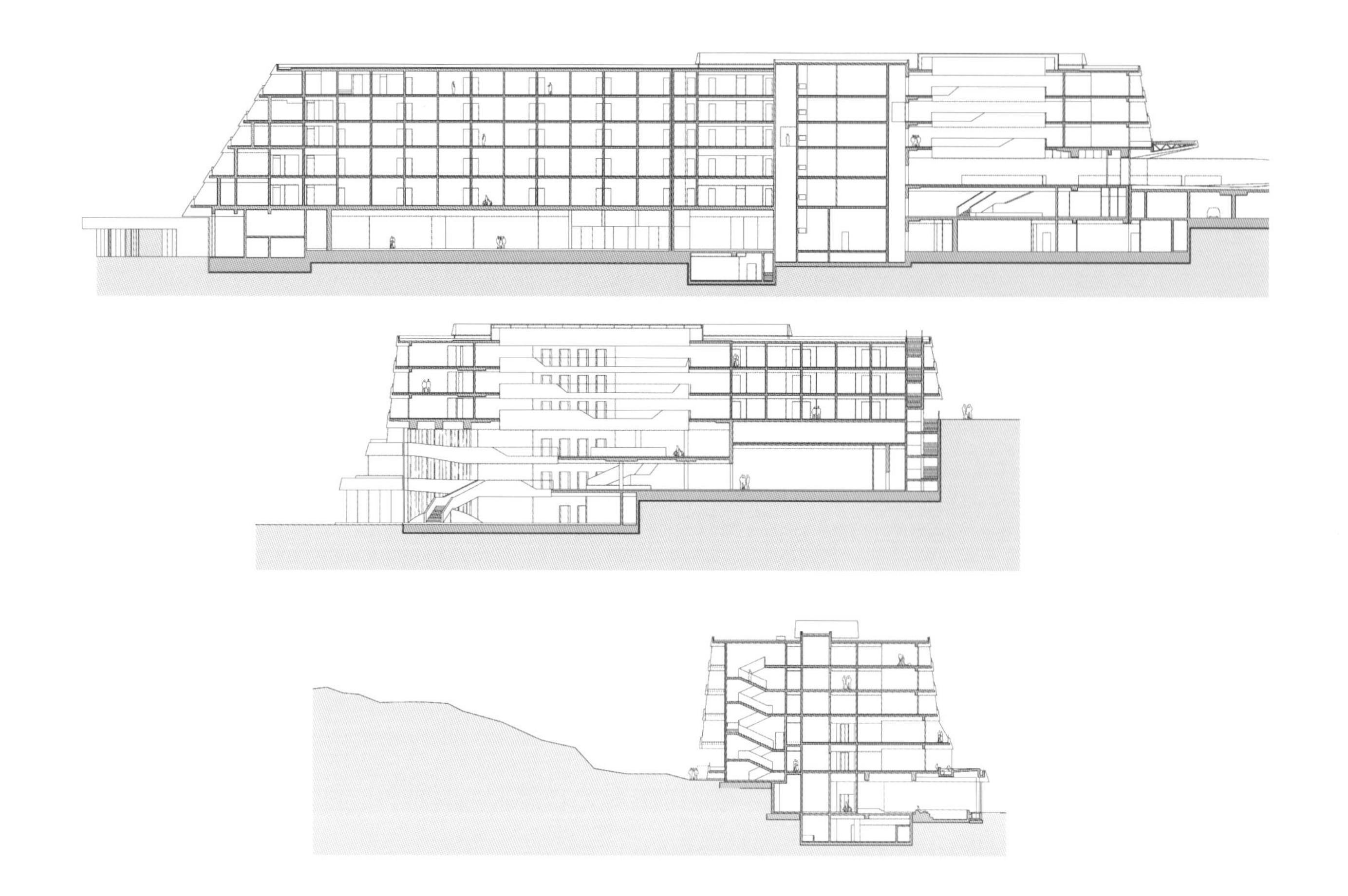

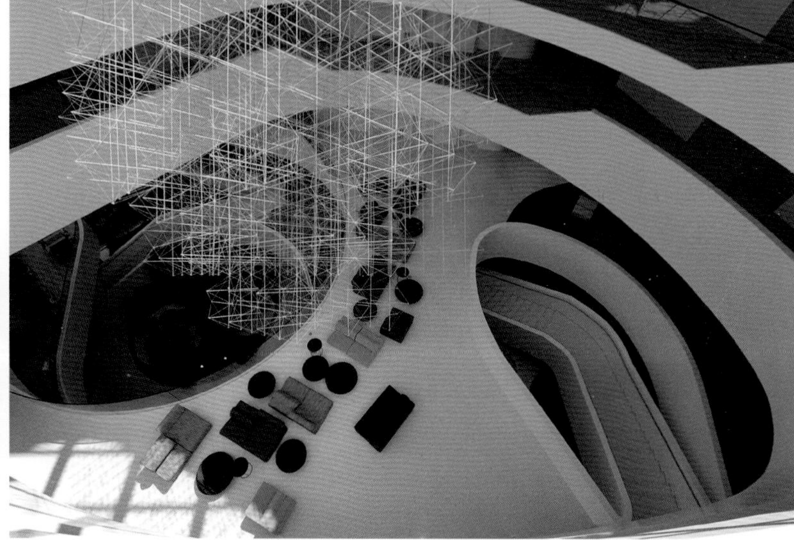

叠式建筑

Office Building ZAC Claude Bernard – Paris XIX

克洛德·贝尔纳办公楼

Design Company / 设计事务所:
ECDM
Location / 地点:
France（法国）
Area / 面积:
12 000 m^2
Photography / 摄影:
Benoît Fougeirol

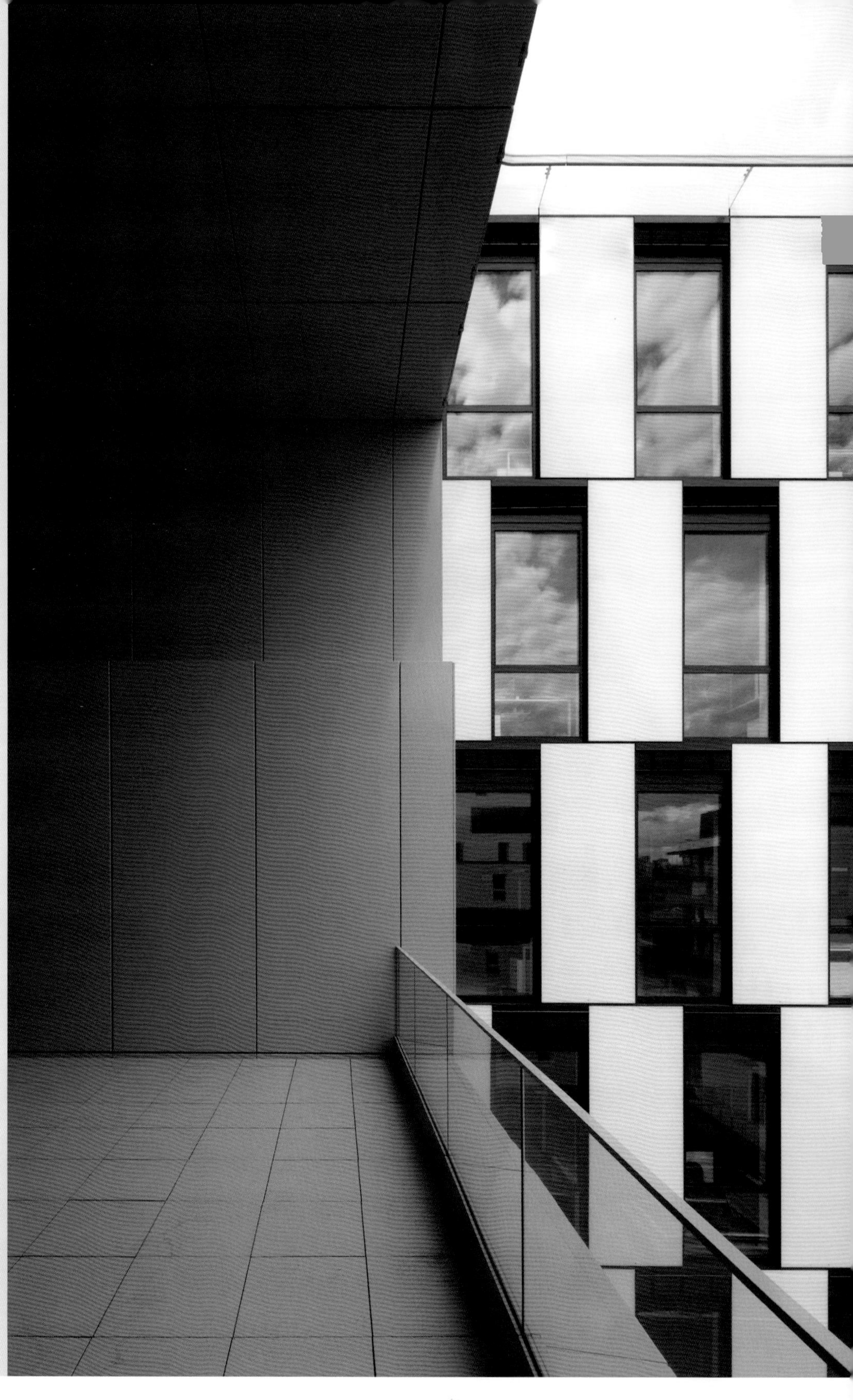

视觉亮点

建筑表皮黑与白的对比，不同楼层之间框架结构的错位，使整座建筑具有强烈的韵律感。

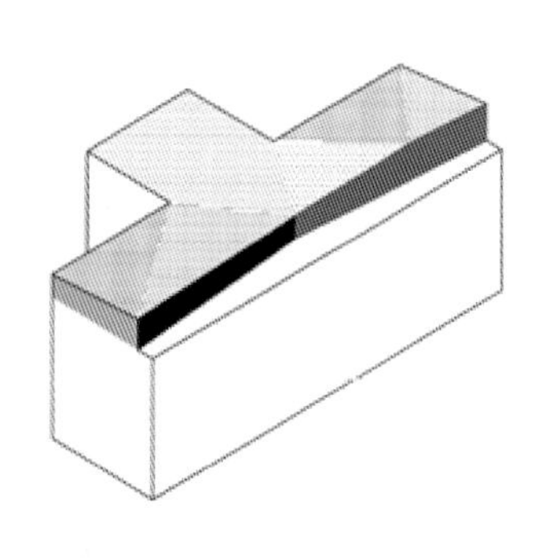

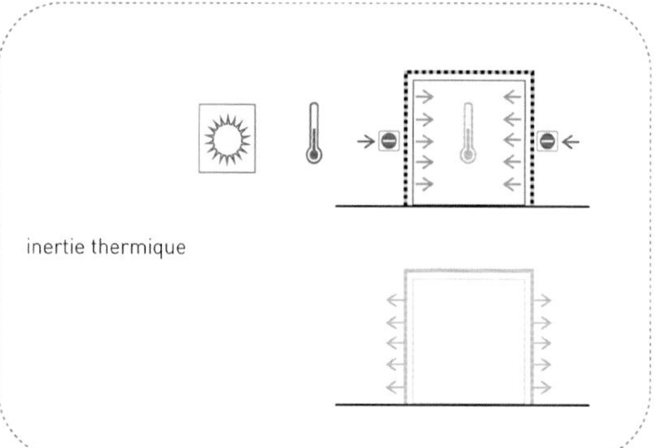

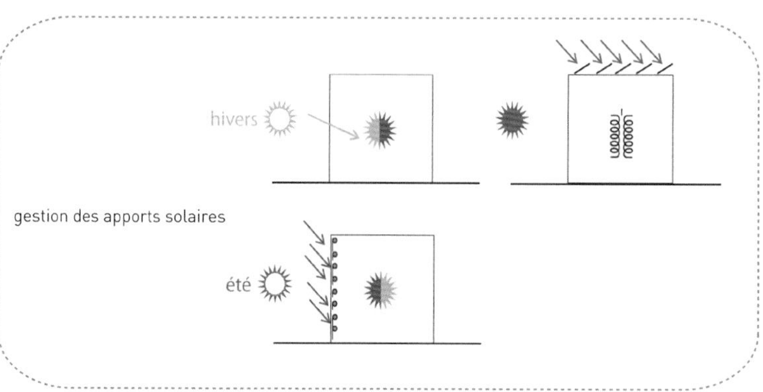

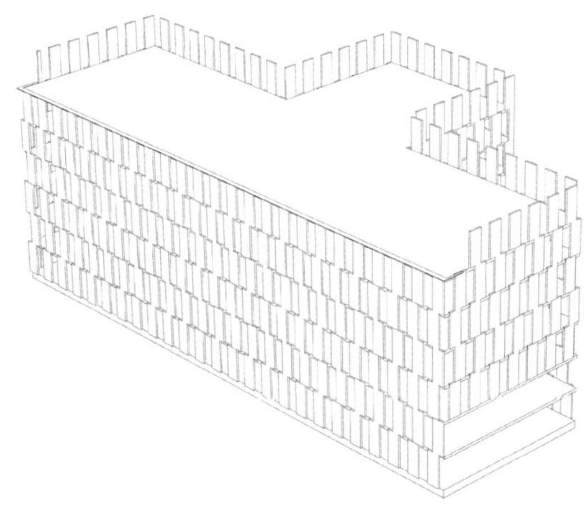

+

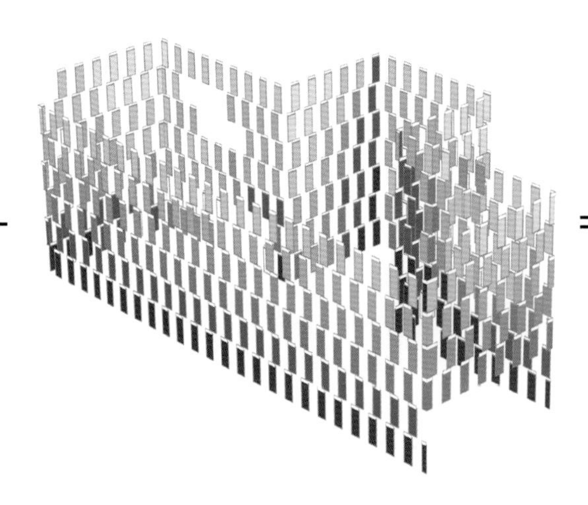

=

forte inertie thermique
+
contrôle des apports solaires

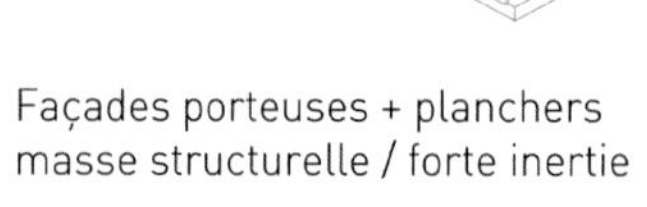

Façades porteuses + planchers
masse structurelle / forte inertie

stores / contrôle de l'apport lumineux

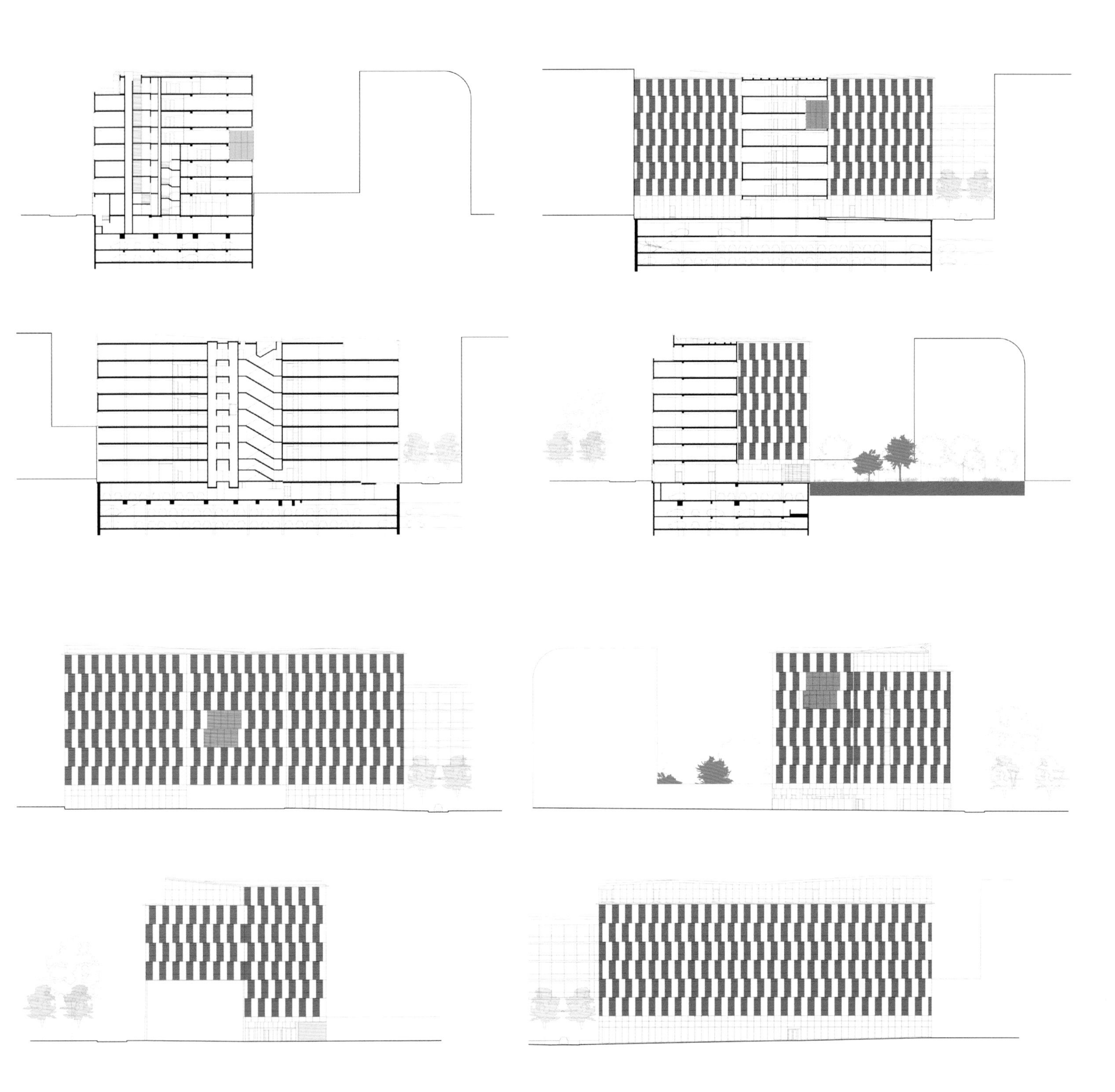

Resulting from work on archetypes volumetric of office buildings, the shape of the project says a simple and clear geometry. In a search for rationality, we propose a tripod organization to distribute all areas around a single central circulation. This device improves the efficiency of the plan in increasing the size of offices while maintaining a limited number of vertical cores. The rationality is reflected in the writing of the facades that expresses strictly a frame of 1.35 m of offices with a rhythmic « full and empty ». The expression of the frame is enhanced by an offset system between levels, which blurs the lines of reading by creating a visual distortion effect, conferring a kinetic aspect to the whole. Rationalism worthy of the great age of « corporate » architecture is confronted with the Venturian ambiguity of the pop architecture.

This pattern runs continuously to surround the building like textured wallpaper. The opaque parts are treated with white enamel grass to make the volume by playing with contrasts, light and reflections. Typical elements of Parisian architecture that is a base, a main building and a penthouse here reinterpreted by the use of a white silkscreened glass accentuates the effect of light.

该项目的建筑外形符合设计师制定的简单、清晰的设计理念。为了使建筑设计更具合理性，设计师提议所有的建筑空间都围绕中心区域展开设计。这样的设计方案提高了空间的利用效率，增加了办公区的面积，同时确保该建筑能拥有固定数目的垂直式的沟通空间。建筑设计的独特之处主要体现在建筑立面上：富有韵律的黑白式框架结构创造了一个新颖、独特的外观。不同楼层之间框架结构稍有错位，营造出一种扭曲式的空间观感，使整个建筑立面更显生动。这样的立面就像织纹式壁纸包裹着整座建筑。该项目完美地呈现出了对比色彩和光影效果。该建筑设计使用一些比较具有代表性的巴黎建筑元素，并通过黑与白的对比进一步营造了强烈的视觉感。

BNP PARIBAS
BNP PARIBAS

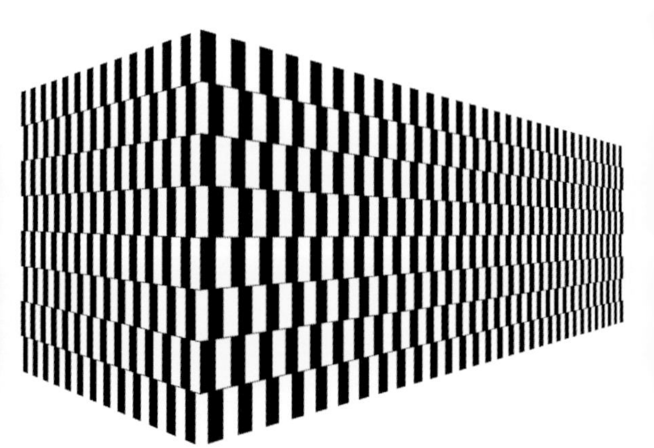

叠式建筑

办公建筑

Stelmat Headquarters

Stelmat新总部大楼

Design Company / 设计事务所:
ABOUT: BLANK ARCHITECTURE
Project Architect / 设计师:
André Lira, Érika Santiago
Location / 地点:
Brazil（巴西）
Area / 面积:
2 500 m²
Computer Renderings / 效果图:
Metrocúbico Digital

视觉亮点

长方形的体块沿不同的角度堆叠，最终创造了一个通透的开放式办公空间。

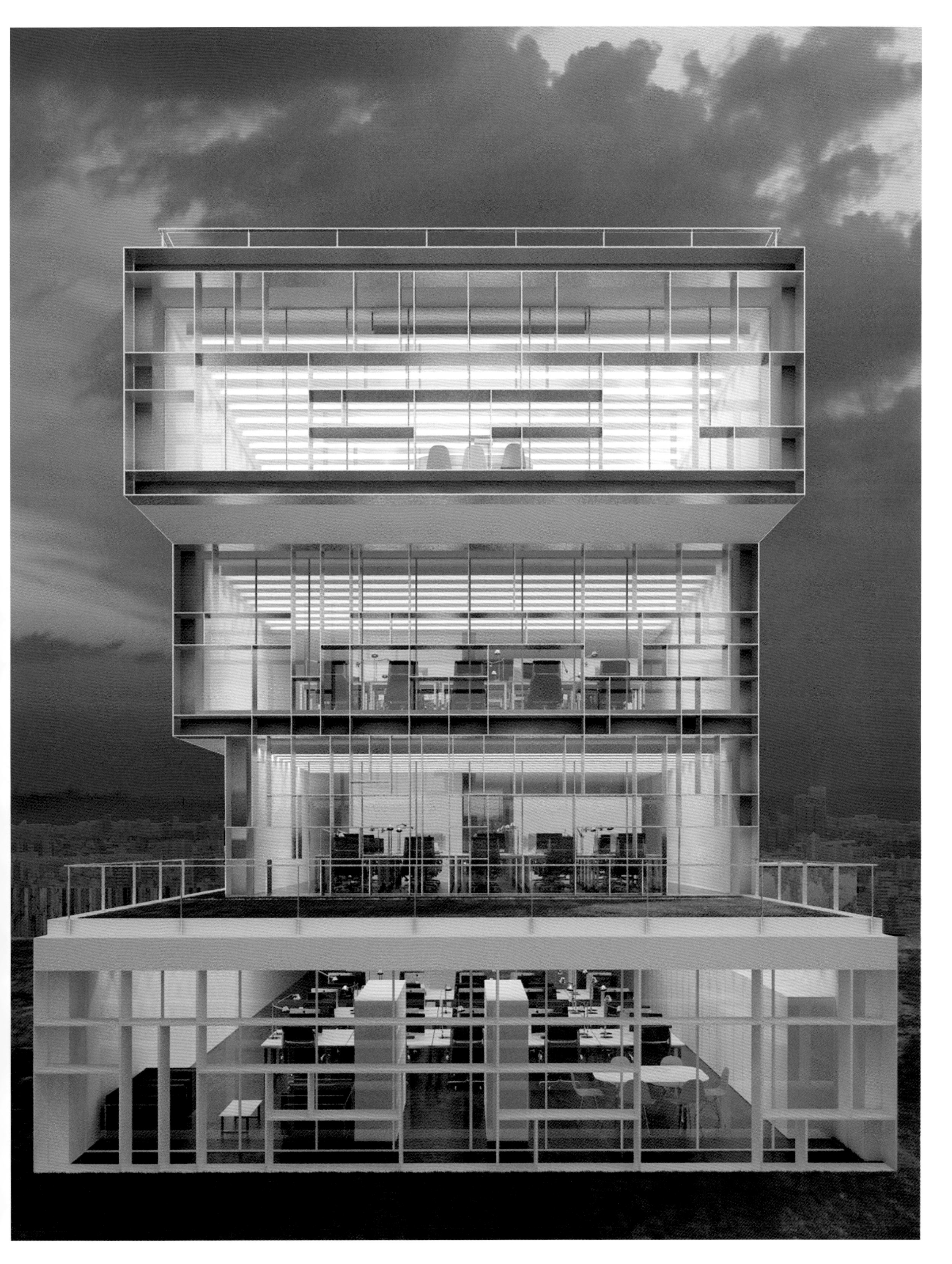

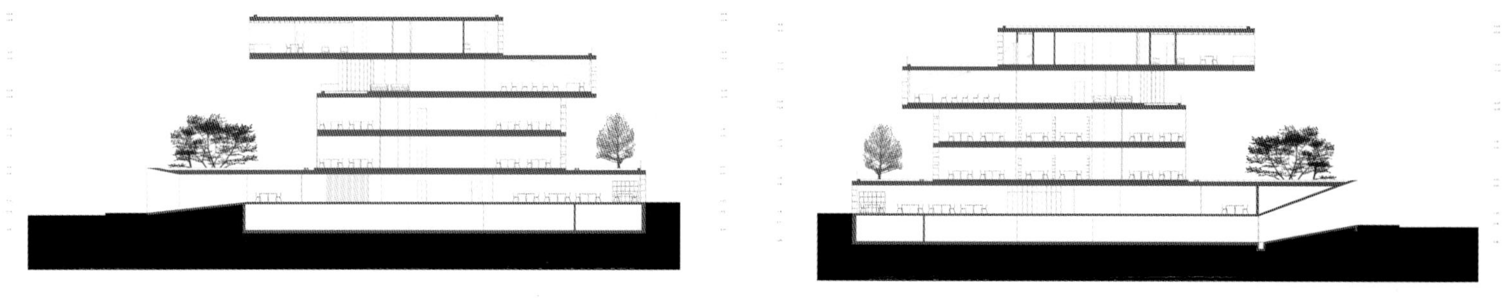

The project for the new headquarters of Stelmat seeks above all a great tune and identification with the company values. The concept rests on two main pillars. The application of the company's maximum, "technology that generates economy" to the building, and exploration of the concept of free marketing inherent to the construction of a new building that stands out in its surroundings. The first pillar is subdivided into two distinct areas: operational technology and technology of sustainability. Both, through the efficiency of the created layouts, use of materials and a rigorous study of sustainability, contribute to create a building that we intend to have an active role in the company's productivity and with influence in its economic performance. The second pillar that consists on the free marketing is based on the capacity it is intended that the building will have, to become a city landmark, with wide repercussions in the dissemination of Stelmat brand and philosophy, not only for customers but also to the general public.
The volume gains motion with the use of the cantilevers, surprising by its balance. The 2 500 m^2 of office area, develops on an open space layout, with a strong relationship with the exterior, and punctuated by external areas with balcony or green roof terraces, contributing to create a better work environment and therefore a more productive company.

Stelmat新总部大楼的主要设计理念是，打造出一座能充分展现公司价值观——拉动经济的技术——的建筑，使这座新总部大楼在周边建筑中脱颖而出。建筑内部分为两个区域：技术运行区和技术可持续区。两者都通过有效的布局、材料的应用和可持续性的建设相互呼应和统一，这两个空间在提升公司生产力水平和经济运行方面发挥了重要作用。此外，该建筑设计的另外一个目标是使其成为城市的地标性建筑，在宣传公司品牌和经营理念方面发挥重要作用。建筑面对的不仅是每个公司的客户，也要给公众留下深刻的印象。
建筑的悬臂式结构实现了惊人的平衡，并赋予整座建筑一种动态的美感。2500 m^2 的办公区域拥有开放式的空间布局，露台和屋顶绿化营造了宜人的室外景观，使建筑内外有着密切的联系，并对改善办公环境、提升工作效率有着很大帮助。

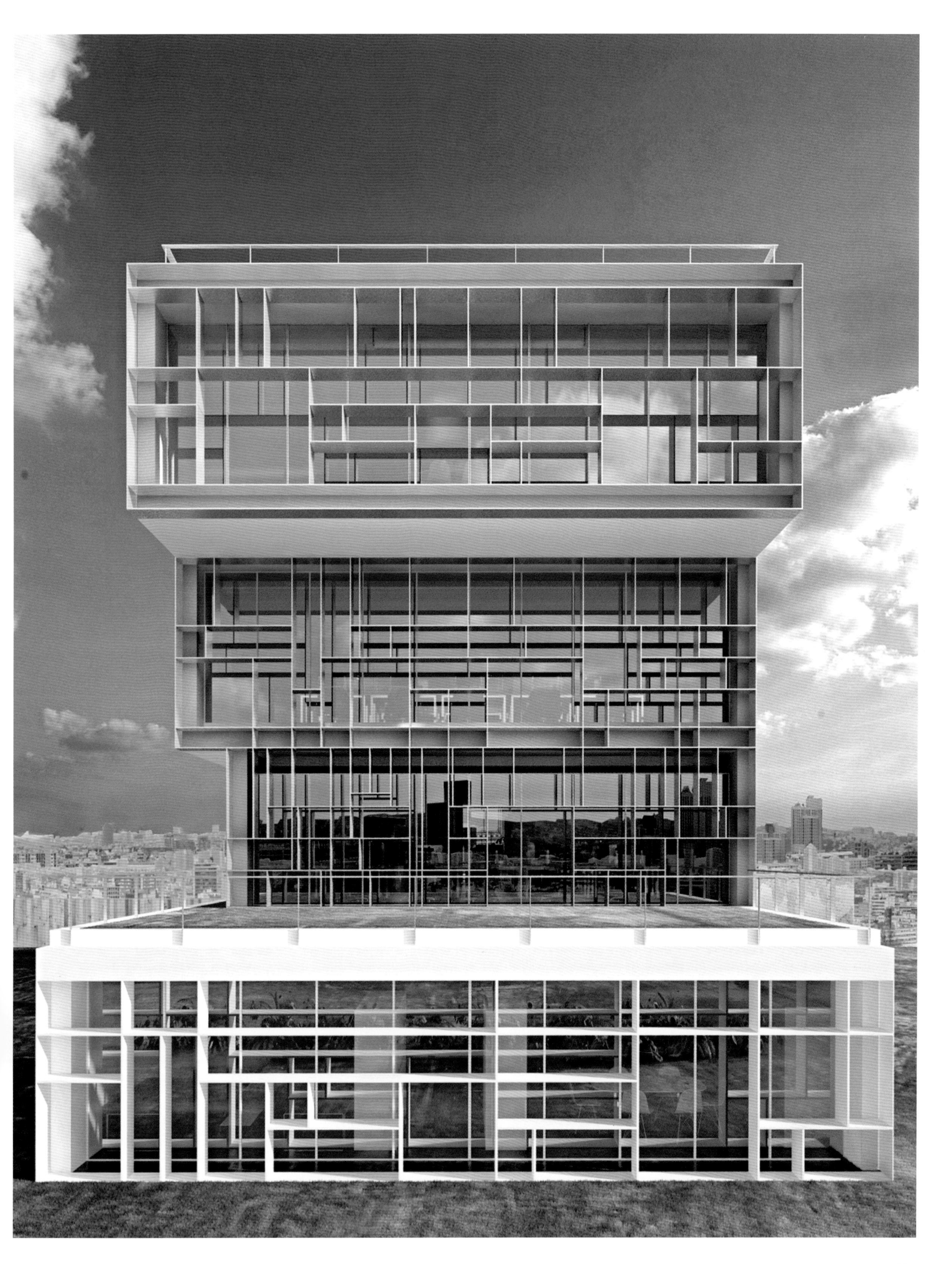

stelmat*

M-MAYBE HE BECAME ILL AND COULDN'T LEAVE THE STUDIO

stelma*

Head Offices of the Telecommunications Market Commission

西班牙电信总部办公楼

Design Company / 设计事务所:
Battle & Roig Architects
Project Architect / 设计师:
Enric Batlle, Joan Roig
Location / 地点:
Spain（西班牙）
Area / 面积:
12000 m^2

视觉亮点

水平金属板材的层叠式应用，不仅构成了建筑的主体结构，还使该建筑与周边建筑相互呼应。

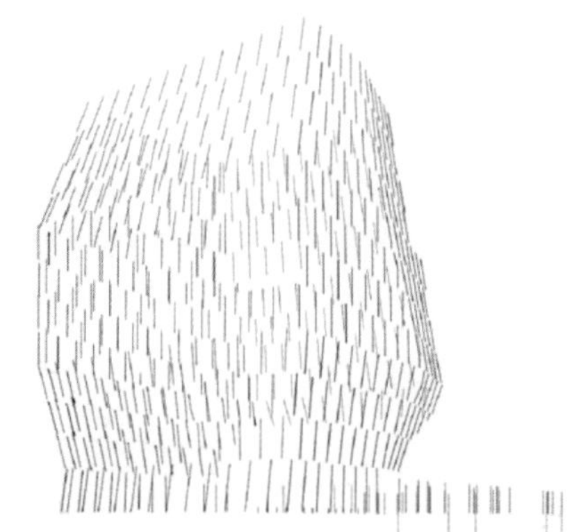

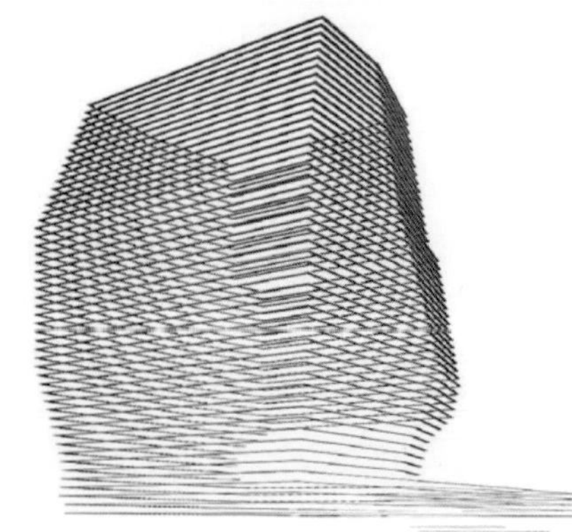

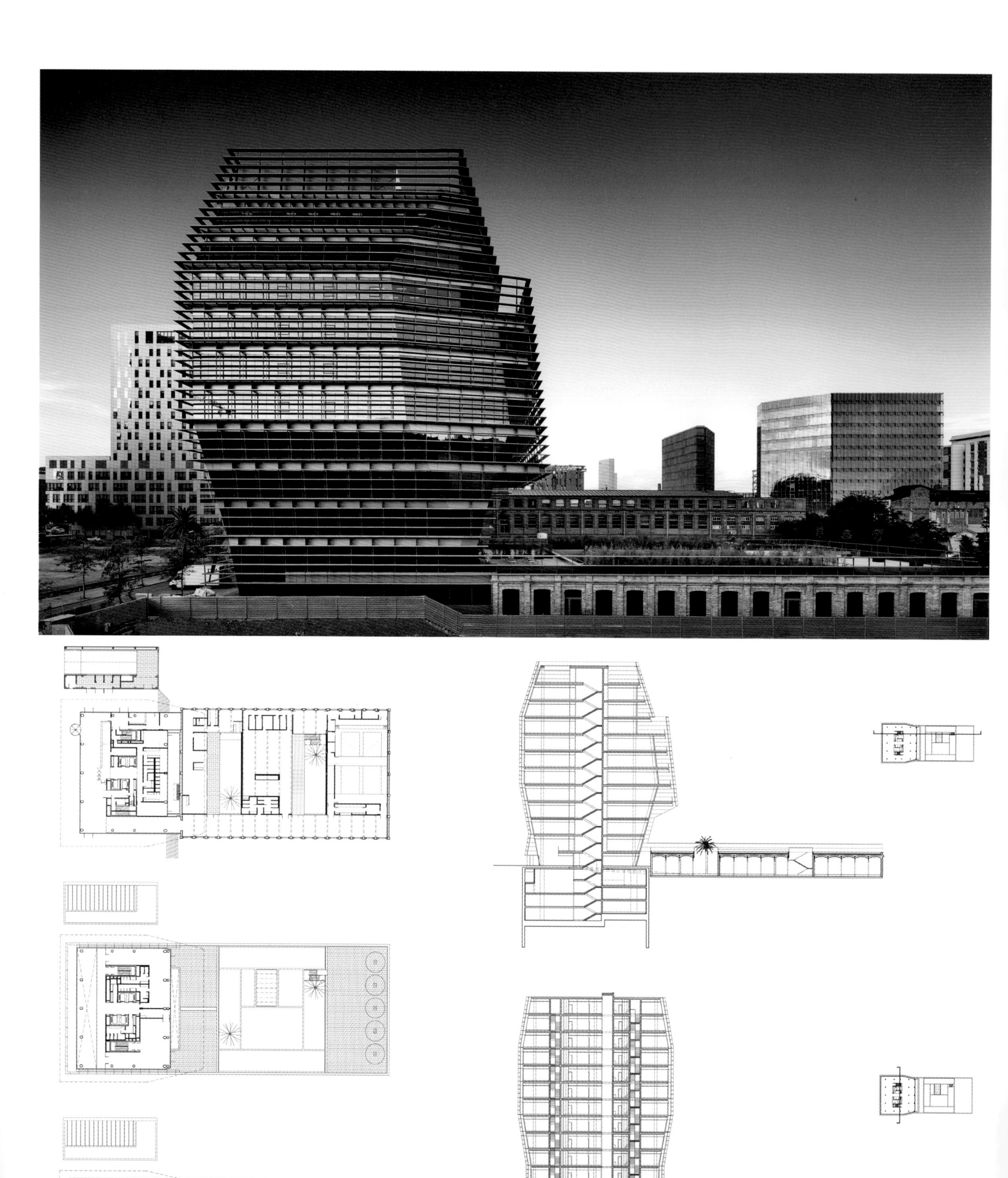

The CMT building stands on a long, narrow site that presents its main façade to Carrer Bolivia and is delimited to either side by a passage. One of the old Can Tiana factory buildings stands at the centre of the site, and the project sets out to recover and incorporate it into the CMT's functional programme.

The main volume, containing the offices, is organized around a central nucleus of entrances and services, and the workstations are laid out around it, making full use of the spatial freedom enjoyed by a building that opens out on all four sides. The variation and superposition of exterior spaces and workspaces serve to direct the volume towards the old factory and establish a subtle, utilitarian relation. The distant presence of the sea and a south-facing orientation determine the correct position of the terraces.

The decision to bring a unitary treatment to the building's outer appearance led us to protect its façade using a horizontal slat system throughout its volume that continues over the old factory, connecting the two. The slats serve to cover the upper terraces and installations, and form an awning at the ground floor entrance.

该总部大楼坐落在一处狭长的地块上，地块周边被几条大街环绕。一座老旧的厂房建筑坐落在地块中央。设计师致力于将这座建筑融入周边的环境中。

设置了办公区的建筑主结构围绕着中央的入口区和服务区展开设计。该建筑沿四个方向向外伸展。独特的层叠式外观使其从周边的建筑中脱颖而出。富于变化的建筑面向老旧的厂房，建筑的颜色、质感都与老旧的厂房有所呼应，远处的大海和建筑的朝向（朝向南方）确定了建筑露台的具体位置。

为了营造建筑外表皮的统一感，设计师采用水平金属板条叠加的方法建造建筑的主结构，这种材质还被用于老厂房的装饰，这样两座建筑就更加融为一体了。这种层叠式结构一直延续至建筑顶部，成为屋顶露台的遮蔽设施。

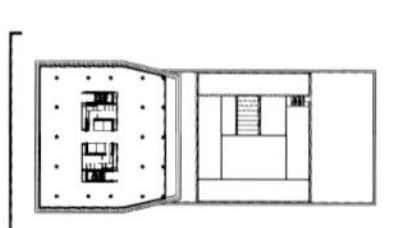

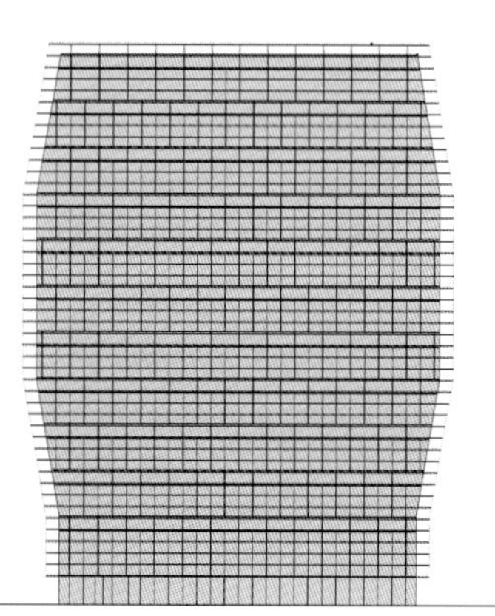

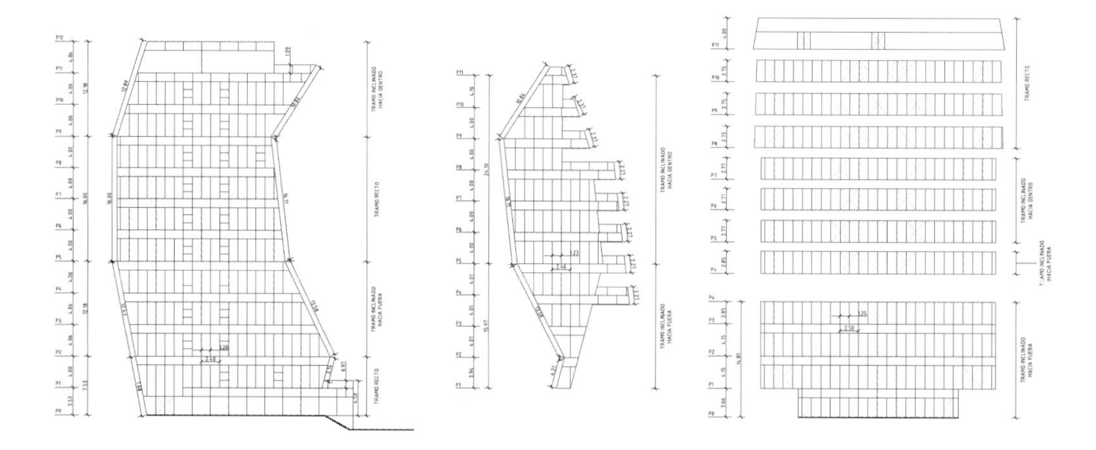
TRAMO INCLINADO HACIA DENTRO
TRAMO RECTO
TRAMO INCLINADO HACIA FUERA
TRAMO RECTO
TRAMO INCLINADO HACIA DENTRO
TRAMO INCLINADO HACIA FUERA
TRAMO RECTO
TRAMO INCLINADO HACIA DENTRO
TRAMO INCLINADO HACIA FUERA
TRAMO INCLINADO HACIA FUERA

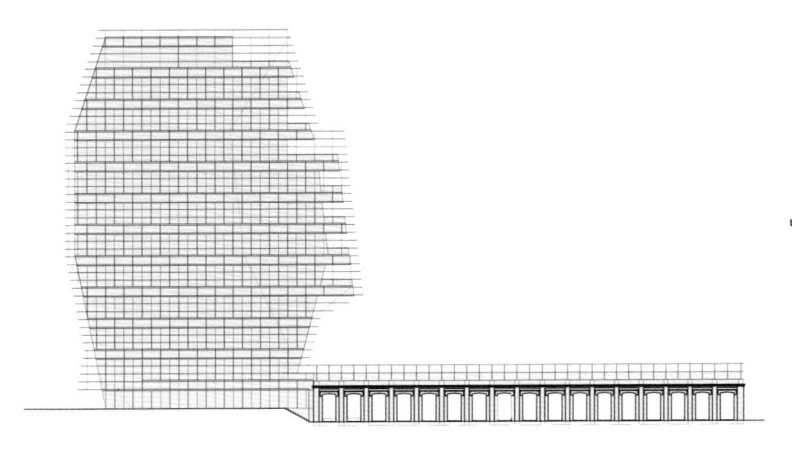

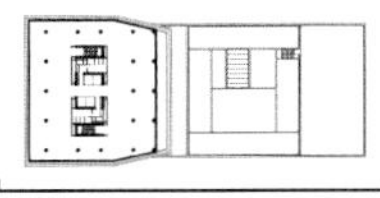

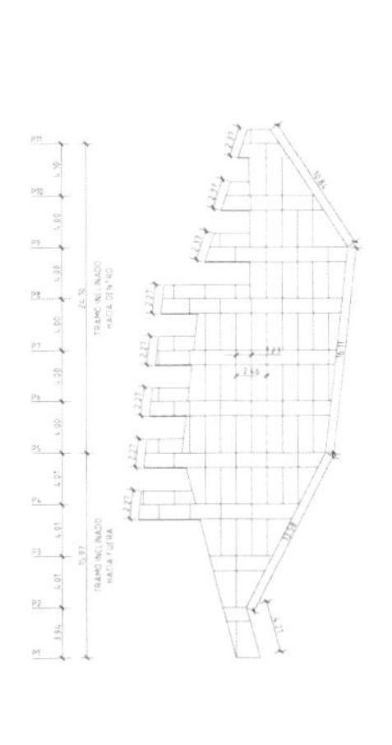

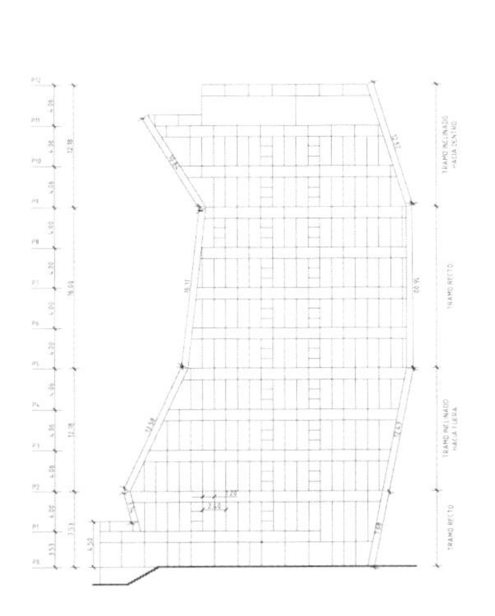

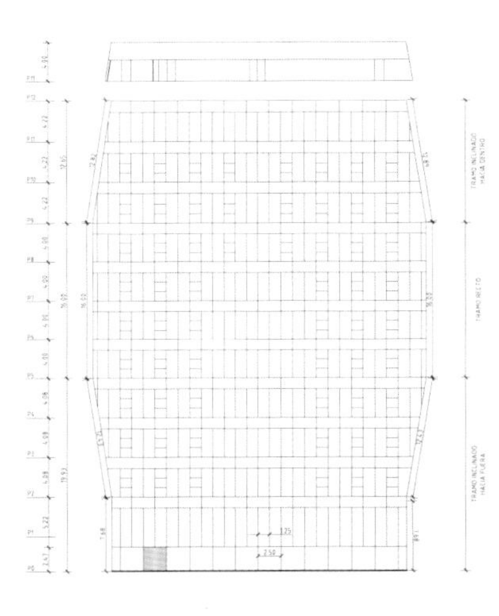

ro / Registre

Basket Apartments in Paris

巴黎巴斯哥特学生公寓

Design Company / 设计事务所:
OFIS arhitekti
Project Architect / 设计师:
Rok Oman, Spela Videcnik
Location / 地点:
France（法国）
Area / 面积:
site 1 981 m^2, building 931 m^2, gross floor area 8 500 m^2, landscape 1 050 m^2
Photography / 摄影:
Tomaz Gregoric

视觉亮点

不同朝向的建筑单元有规律、有秩序地叠加起来，不仅赋予每个结构单元独特的观景视角，而且使建筑看起来充满变化和动感。

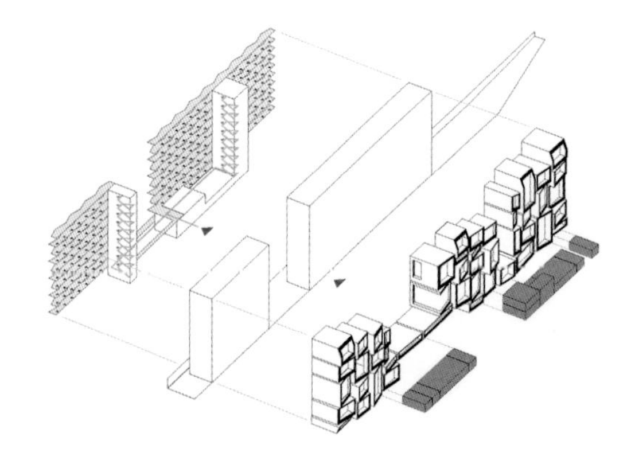

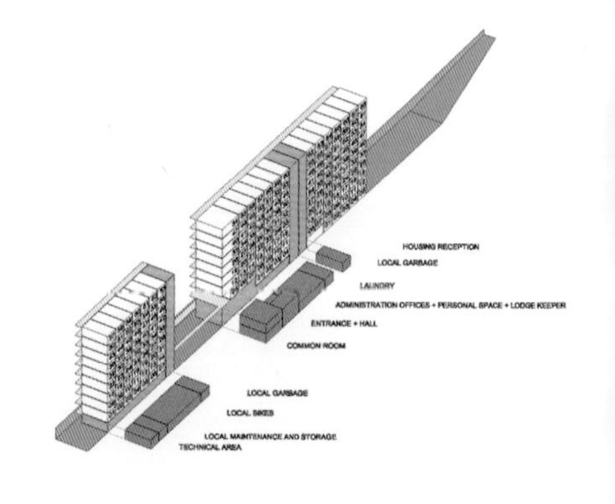

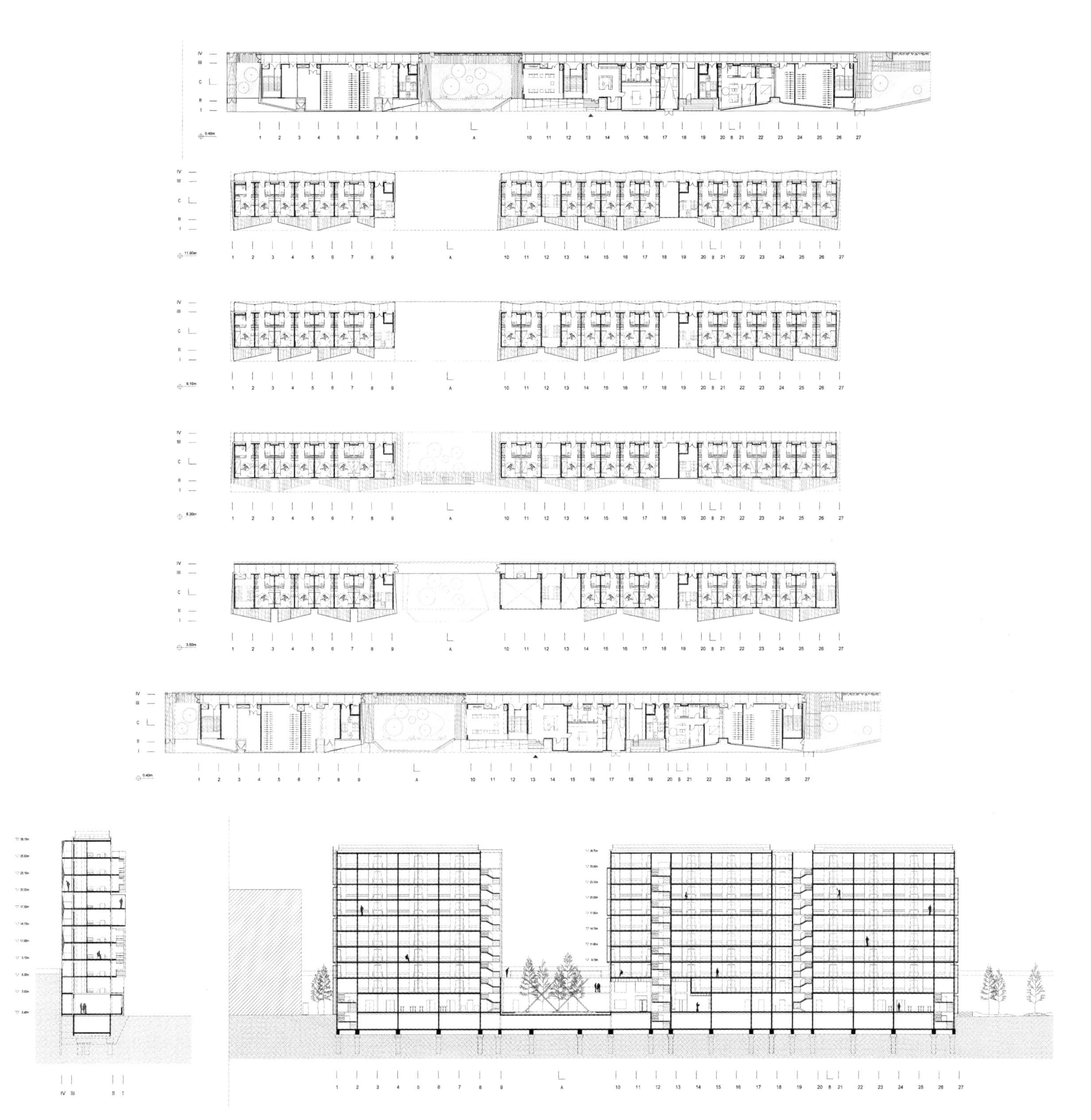

The parcel has a very particular configuration; 11m in width and extending approximately 200m north-south. This foreshadows the importance of processing the eastern facade overlooking the extension of the street Des Petits Ponts which hosts the tram and both cyclist and pedestrian walkways.
Aarrow length of the plot with 10 floors gives to site a significant presence. Each volume contains two different faces according to the function and program: The elevation towards the street Ddes Petits Ponts contains studio balconies-baskets of different sizes made from HPL timber stripes. They are randomly oriented to diversify the views and rhythm of the façade. Shifted baskets create a dynamic surface while also breaking down the scale and proportion of the building.
The elevation towards the football field has an open passage walkway with studio entrances enclosed with a 3D metal mesh. Both volumes are connected on the first floor with a narrow bridge which is also an open common space for students.

该项目具有非常别致的空间设计，整座建筑宽 11 m，呈南北走向，长约 200 m。项目所处的地块为狭长形，建筑周边的街道上既设有电车轨道，也设有自行车道和人行道。
矗立在狭长地块上的这座十层建筑的外观相当醒目。每处结构都拥有两个功能朝向街道的外立面上设置了可以俯瞰街景的不同规模的阳台，阳台使用 HPL 纹状木材打造而成；阳台的朝向角度各不相同，使人们可以多角度地欣赏美景，同时赋予建筑立面节奏感。多变的结构打造出了富有活力的建筑立面。
朝向足球场一侧的立面拥有开放式的通道，可通向各间公寓，公寓入口处使用三维立体金属网进行装饰。建筑一层设置了一座连接两栋公寓的天桥，这也是供学生们使用的一处开放式公共空间。

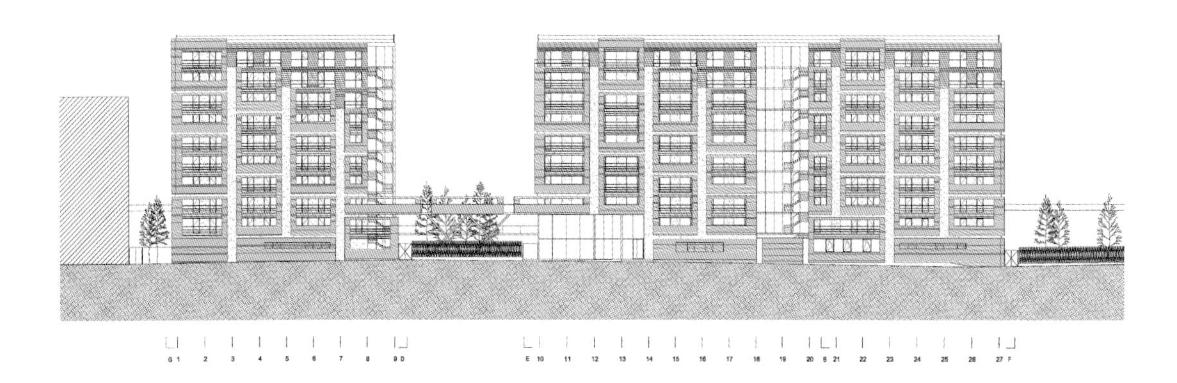

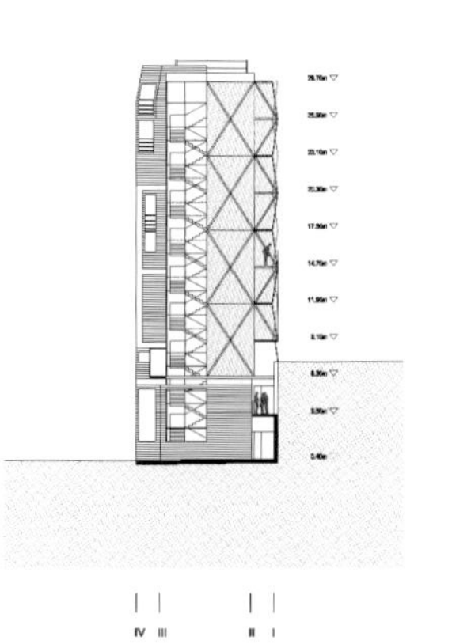

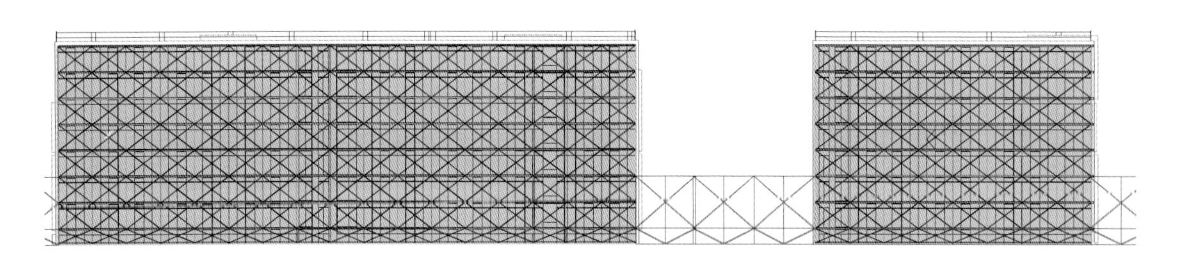

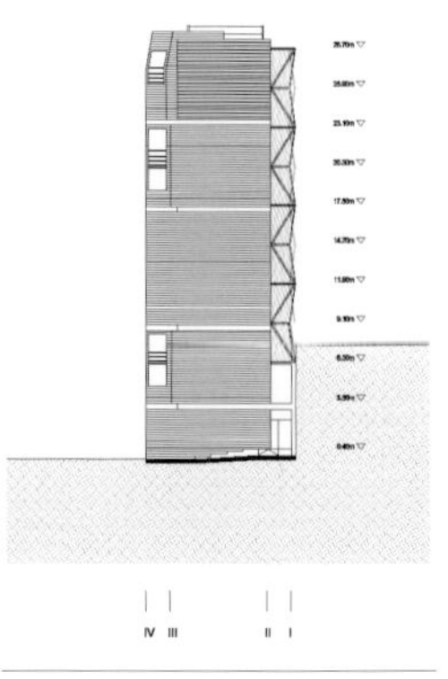

B
08

叠式建筑

居住建筑

Berge du Lac

贝格迪拉克街区住宅

Design Company / 设计事务所:
Nicolas Laisné* Architecte Urbaniste,
CHRISTOPHE ROUSSELLE architecte urbaniste

Project Architect / 设计师:
Nicolas Laisné, Christophe Rousselle

Location / 地点:
France（法国）

Area / 面积:
8 100 m^2

Photography / 摄影:
Cyrille Weiner

视觉亮点

建筑轮廓与建筑周边电车轨道形状相呼应。

This project, located in the new Berge du Lac area in Bordeaux, includes four collective buildings and ten individual houses which form a block. The dark shades and unbalanced silhouettes of the three tallest structures echo the tramway. On the other side, the individual houses and a building, covered in wood, are erected along the side of a linear park. Each of these architectural choices responds to the particular circumstances of these buildings.

该项目位于波尔多新建的贝格迪拉克区，这是由四座公共建筑和十座独立住宅建筑形成的一处街区。

其中三座最高建筑形成的昏暗阴影和不规则轮廓与电车轨道遥相呼应。另一方面，外部覆以木材的独立式住宅和一座公共建筑位于呈线性设置的公园的一侧。每一个建筑设计理念都充分考虑了这些建筑周边的特殊地理条件。

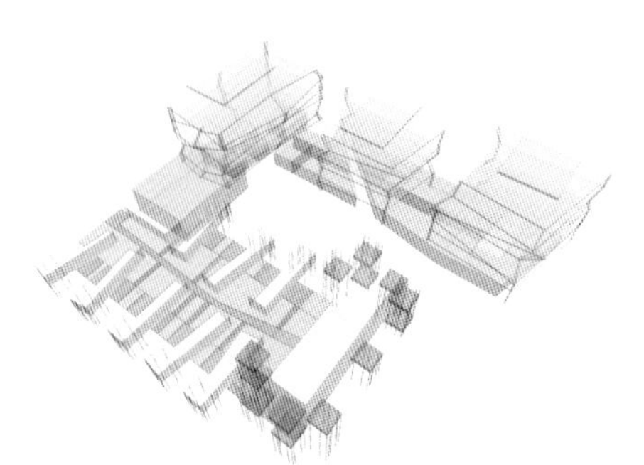

A vast garden in open ground

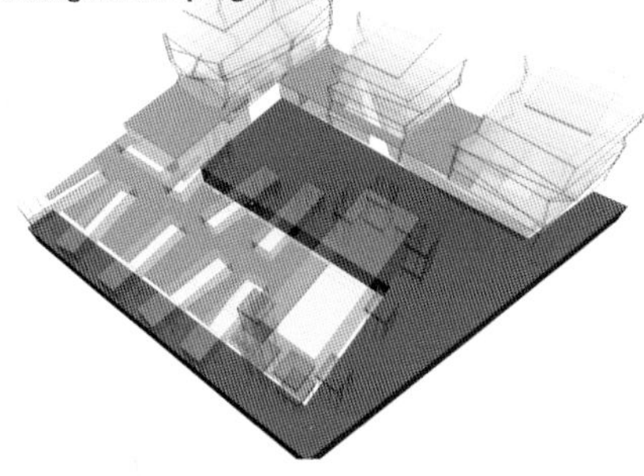

Double height hall

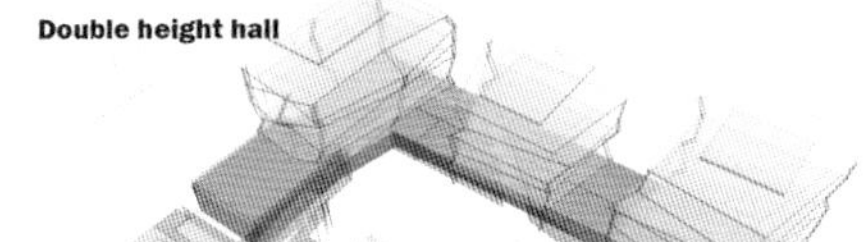

Mixed-use single-storey house with the tramway court and a garden

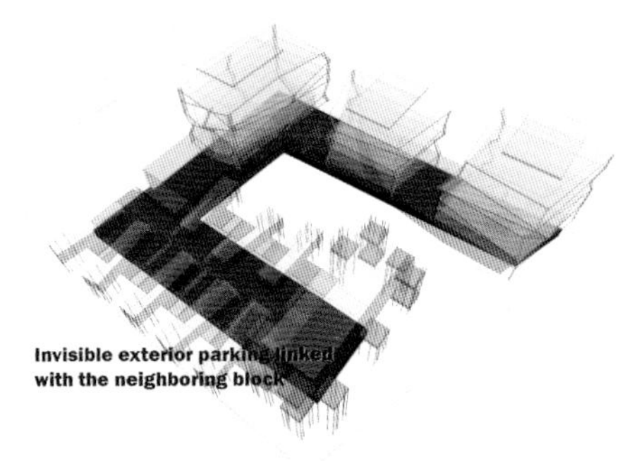

Invisible exterior parking linked with the neighboring block

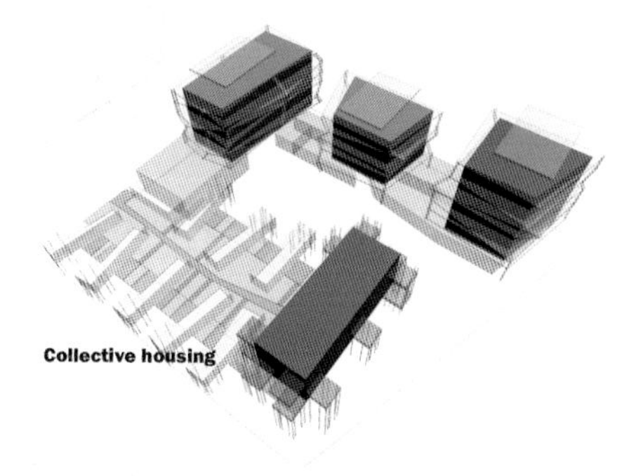

Collective housing

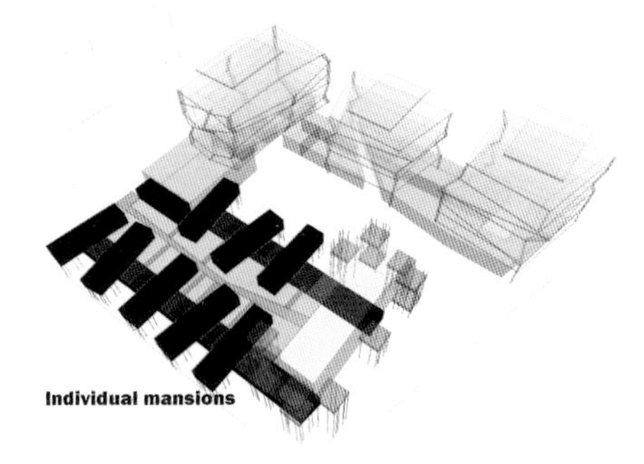

Individual mansions

Office Building in Pujades 22@

普亚达斯办公楼

Design Company / 设计事务所:
Josep Miàs
Project Team / 设计团队:
Adriana Porta (project leader), Carles Bou (technical advisor), Silvia Brandi, Dani Montes, Baptiste Marconnet, Mario Blanco, Diogo Henriques
Location / 地点:
Spain （西班牙）
Area / 面积:
7 000 m^2
Photography / 摄影:
Jordi Bernadó

视觉亮点

建筑的建造过程就像孩童搭积木一样，长方形的体块被一层层建造起来。建筑整齐、简洁，每一个体块都是一个独立的空间。

COMARCAS CATALANAS, S. A.

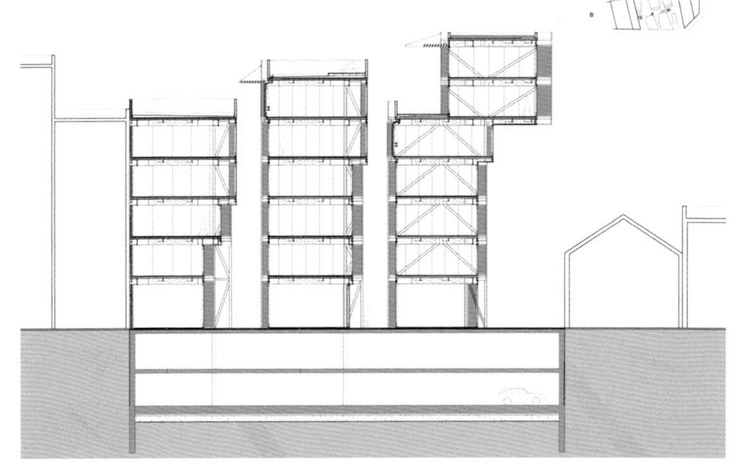

The aim of the project is to rebuild from the old plot division rules, following a geometry that obeys to Pere IV Street orientation. The elevations of the building show clearly how these transversal secondary streets appear between the three structures, as if they were passages, and allow to illuminate the interior spaces of the offices.
Consequently, the geometry does not maintain a constant alignment. It presents some doubts about the section in relation to the passage or the street-courtyard and in relation to the existing industrial building in Pamplona Street as well. That's the reason why the building reacts and overflies the old construction on the upper floors. The project in 22@ presents a fragmented building with vertical voids between the volumes. Underground, the building presents a two-floor parking with access next to the preserved industrial building. The structural system is a steel skeleton that presents several levels of complexity. The building is proposed as six independent entities joined by a transversal central passage where stairs and elevators are located.

该项目的主要设计目的是改变原有的地块形象，建筑的平面布局要与建筑所处地块周边街道的走向相适应。该项目设计团队最终在普亚达斯大街和佩雷第四大街之间打造了三座建筑结构。
这些建筑结构既互相联系，又有各自的变化。设计师要处理好办公建筑与周边街道及潘普洛纳大街上原有的工业建筑之间的关系，设计师的应对策略是提升新建的办公楼的高度，使其高于周边的工业建筑。该项目为叠式建筑，建筑内部空间十分通透。
该建筑拥有两层的地下停车场，其入口通道靠近原有的工业建筑。建筑为钢结构。建筑的各个独立部分通过横向通道连接在一起。

图书在版编目(CIP)数据

全球建筑设计风潮.2，条形 叠式建筑 / ThinkArchit工作室主编. —武汉：华中科技大学出版社，2013.11
ISBN 978-7-5609-9340-9

Ⅰ. ①全… Ⅱ. ①T… Ⅲ. ①建筑设计—作品集—世界—现代 Ⅳ. ①TU206

中国版本图书馆CIP数据核字(2013)第193436号

全球建筑设计风潮2 条形 叠式建筑

ThinkArchit工作室 主编

出版发行：华中科技大学出版社（中国·武汉）
地　　址：武汉市武昌珞喻路1037号（邮编:430074）
出 版 人：阮海洪

责任编辑：刘锐桢　　责任监印：秦　英
责任校对：杨　睿　　装帧设计：张　宇

印　　刷：小森印刷（北京）有限公司
开　　本：965 mm × 1230 mm　1/16
印　　张：19
字　　数：160千字
版　　次：2013年11月第1版第1次印刷
定　　价：318.00元（USD 76.99）

投稿热线：(010)64155588-8009　hzjztg@163.com
本书若有印装质量问题，请向出版社营销中心调换
全国免费服务热线：400-6679-118 竭诚为您服务